Biology of Evolution and Systematics

i

BIOLOGY OF EVOLUTION AND SYSTEMATICS

A Cohesive, Concise, yet Comprehensive

Introduction

for Students and Professionals

Paul Sanghera, Ph.D.

Biology of Evolution and Systematics: A Cohesive, Concise, yet Comprehensive Introduction for Students and Professionals

Published by
Infonential, Inc.

Technical editor: Dr. John Serri

Copy editor: Mary C. Moore

ISBN-13: 13: 978-1514782002
ISBN-10: 1514782006

To:

Biology students and professionals
all across the world

About the Author

Dr. Paul Sanghera is an educator, scientist, engineer, technologist, author, and entrepreneur. He has a diverse background and experience in multiple fields including physics, chemistry, mathematics, computer science, and biosciences. He holds a Ph.D. in Physics from Carleton University, Canada; a Master of Engineering degree in Computer Science from Cornell University, U.S.A; and a B.Sc. with a triple major in physics, chemistry, and math from G.N.D. University, India. He has comprehensive, cross-disciplinary, cross-continental experience in research, teaching, and learning. He has taught a wide spectrum of science and technology courses all across the world, from San Jose State University, U.S.A. to the Carleton University and Simon Fraser University, Canada; to the Indian Institute of Technology (IIT), India. He has authored and co-authored about 150 scientific research papers on the subatomic particles of matter published in well-reputed European, American, and international research journals. At world class laboratories, such as CERN in Europe and Nuclear Lab at Cornell, he has participated in designing and conducting experiments to test the quantum theories and models of subatomic particles, and thereby contributed to testing the Standard Model of the universe.

Dr. Sanghera was a lead engineer at the first Internet (Web) company Netscape Communications Inc. He has been at the ground floor of several startups in Silicon Valley, California and elsewhere; and he has contributed to the development of state-of-the art technologies such as Novell's NDS, the first computer network management system; and one of the first commercial web storefronts at Weborder Inc.

Dr. Sanghera is the author of almost two dozen books in science, technology, and project management. His current research interests involve topics in biomolecular engineering and bioinformatics.

About the Book

Biology of Evolution and Systematics: A Cohesive, Concise, yet Comprehensive Introduction

In this book, Dr. Paul Sanghera, the bestselling author of several books in science and technology, provides a cohesive, concise, yet comprehensive coverage of the key concepts of evolution and systematics in an accessible way. The book presents material in a logical learning sequence: each section builds upon previous sections and each chapter upon previous chapters. All concepts—simple or complex—are well-defined and clearly explained the first time they appear. There is no hopping from topic to topic and no technical jargon without explanation. This book is useful for both students and professionals in biology. Students can use the distilled information in this book to excel in their assignments and exams including AP Biology. Even though this book is self-contained, it also works as a great supplement to any textbook in general biology. Professionals in a biology-related field can use it as a quick reference guide or for a concise review of fundamental concepts, whereas the newcomers can use it as their gateway into the field to swiftly ramp up to speed.

The chapters in the book have the following special features:

- **Note:** A *Note* is used to present additional helpful material related to the topic being described or to emphasize a concept.

- **Caution:** A *Caution* is used to highlight a point which either is crucial or may not fit into a framework of common sense.

- **Think About It:** This feature presents questions or simple problems with answers and solutions to emphasize critical concepts.

- **Problems:** Problems are presented with solutions to explain mathematical concepts.

- **Review Questions:** Review questions with answers are presented at the end of each chapter in order to enable you to test your knowledge and detect your strengths and weakness.

- **Glossary:** This feature permits straightforward access to key terms.

Enjoy!

Contents

Chapter 1

Overview of Evolution

1.1 Evolution of Life: The Big Picture

The big picture of life on our planet can be expressed in three words: unity, diversity, and adaptation. Putting it into one sentence; there is an underlying unity behind the vast diversity of life forms through shared characteristics (traits), which are adapted to their respective environments in remarkable ways. Any reasonable theory of life on earth must explain this big picture: unity, diversity, and adaptation.

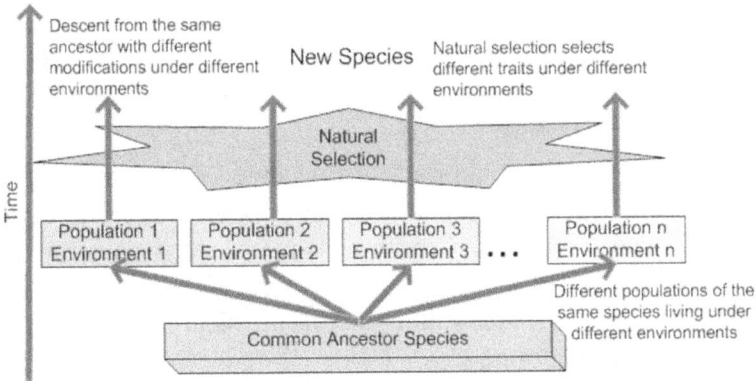

Figure 1.1 **The core concept of evolution: descent with modification**

As illustrated in Figure 1.1, the theory of biological evolution explains how the big picture of life can also be

Biology of Evolution and Systematics, by Paul Sanghera
Copyright © 2015 Infonential.

equivalently expressed in three words: common ancestor, descent, and modification; or into one phrase: descent with modification from a common ancestor. Putting it into one sentence; diverse species, which are descended from a common ancestral species, have modifications facilitated by the process of natural selection. Descent with modification gave rise to diversity, descent from a common ancestor remained the source of the underlying unity behind diversity, and natural selection facilitated the accumulation of diverse sets of modifications that fit the specific environments in which different groups of organisms lived generations after generations. These modifications are dependent on the surrounding environment of each group of organisms. Each unique environment creates different modifications on the life forms. These modifications are called adaptations.

> Note. The phrase *descent with modification* was used by Charles Darwin to explain the big picture of life in his book: *The Origin of Species*.

Figure 1.2 illustrates the course of evolution from genetic mutations to species:

1. **Mutations:** A mutation is a change in the nucleotide sequence of an organism's DNA (deoxyribonucleic acid) that may result from multiple sources such as errors in replication of DNA during cell division, exposure to radiation, and interactions with physical and chemical agents, called mutagens.

2. **Alleles:** Mutations may give rise to new genes and different versions of a single gene. Different versions of the same gene are called alleles.

3. **Genotype variations:** Existence of multiple alleles for a gene gives rise to variations in the genotype (genetic make-up or set of alleles for an organism) across organisms.

4. **Phenotype variations:** Variations in the genotype express themselves as variations in the phenotypes. The phenotypes are the visible physical, anatomical, and physiological traits, among organisms. For example, some of us have brown eyes and some of us have blue eyes.

5. **Natural selection:** Organisms that have *favorable traits* in a given environment tend to survive and reproduce viable offspring at a higher rate than those organisms that do not have favorable traits. Favorable traits are those trait variants that increase the fitness of organisms with their environment. As a result, over generations, favorable traits become dominant and more prevalent in the population, and the population evolves. This process is called natural selection, which operates on phenotype variants.

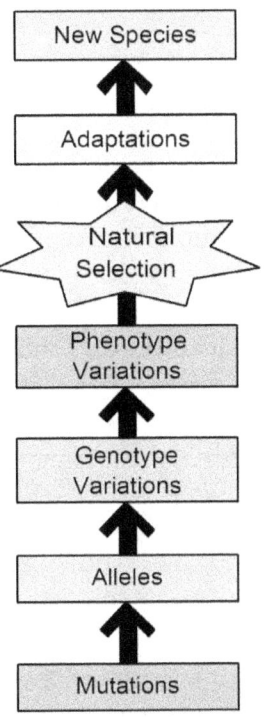

Figure 1.2 The big picture of evolution

6. **Adaptation:** Natural selection selects organisms with favorable traits within a given environment. These inherited characteristics enhance the probability of an organism's

3

survival and ability to reproduce viable offspring. Such inherited characteristics are called adaptations.

7. **New species:** Natural selection, over multiple generations, results in adaptations different from those in previous generations, changing the genetic makeup of the population. If this change is significant enough, this may give rise to a new species.

> **Caution!** It is the organisms that are selected, and as a result of this selection, it is the population that evolves.

We have used the terms species and populations in this section. Let us define these terms. A **species** is a group of organisms that have the potential to interbreed and produce viable, fertile offspring; but do not have the same potential to produce viable, fertile offspring with the members outside of this group. A species may consist of one or more populations. A **population** is a group of organisms of the same species that live in the same area at the same time.

1.2 Evolution of Evolution: History of Evolutionary Thought

The concept of biological evolution was developed over centuries by many natural philosophers and scientists. Although Charles Darwin is the most famous evolutionary biologist, it is also important to realize that he was neither the first to introduce the concept of evolution, nor was he the first to present a mechanism for it. Furthermore, when he did present the mechanism for evolution, called natural selection, he did it at the same time as another scientist named Alfred Russel

Wallace. To understand the study of evolution one must go far back in history.

Greek philosophers: The concept of evolution is traced back to Anaximander (610 BC – 546 BC), a friend and student of Thales who is regarded by many as the father of western philosophy and science. Anaximander proposed that humans must have gradually evolved in the protection of and from other animals. His main argument for this proposal was that as an independent species, human infants would have never survived on their own. However, another famous philosopher Aristotle (384 BC – 322 BC) disagreed. Aristotle who influenced early science in the West more than anyone else proposed that species (life forms)

Figure 1.3 *Scala Naturae*, 1994, by Mark Dion.

were permanent (not changing) and can be arranged on a scale of increasing complexity called *scala naturae*. It is like a ladder (Figure 1.3) with each rung allotted permanently to a specific form of life (say species).

Aristotle's ideas were consistent with creationism, the Old Testament account of creation: species were individually designed and created independent of one another by a supreme being, God, and therefore were perfect and permanent, that is, not changing. Before the beginning of nineteenth century, creationism had been the dominant doctrine even among scientists.

Carolus Linnaeus (1707-1778): Recognizing the similarities among organisms, Carolus Linnaeus, a Swedish botanist and physician, introduced the nested classification system called *taxonomy* as opposed to the linear hierarchy of creationism or the linear *scala naturae* of Aristotle. In taxonomy, all organisms can be classified into interrelated nested hierarchical groups, groups subordinate to groups from top to bottom: kingdom, phylum, class, order, family, genus, and species. This concept is illustrated in Figure 1.4 by using a simple example.

According to this figure, both species *Homo sapiens* and *Homo erectus* belong to the *Genus* Homo, which belongs to the *Family* Hominidae, which belongs to the *Order* Primates,

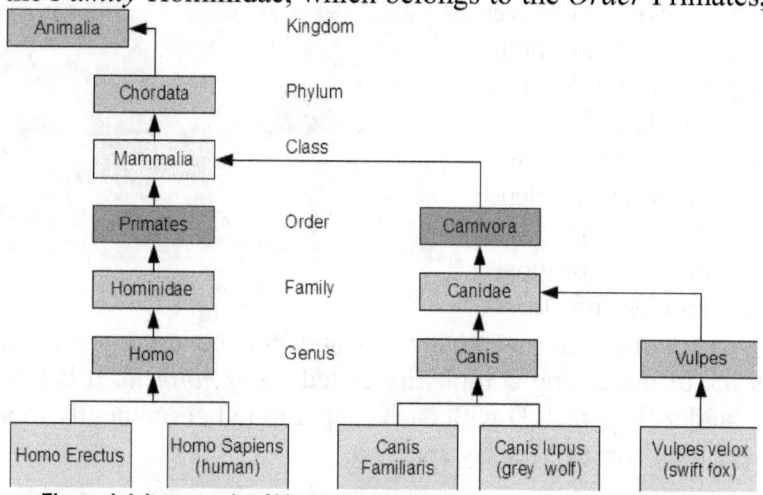

Figure 1.4 An example of Linnaean taxonomy.

which belongs to the *Class* Mammalia, which belongs to the *Phylum* Chordata, and which belongs to the *Kingdom* Animalia. The figure also illustrates that the wolf and the fox belong to the same class (Mammalia) that humans do. Although Linnaeus, known as the father of taxonomy, broke away from the linear hierarchy of creationism, he failed to link his tree (hierarchical) thinking to evolution, probably due to his belief in creationism.

Georges Cuvier (1769-1832): Georges Cuvier, a French scientist, largely founded the field of paleontology, the study of fossils. He observed the following pattern in fossils:

- The fossils on rocks were in form of layers, called stratum.

- The newer the layer of fossils, the closer it was in similarity to the forms of life currently living on earth.

- From one layer to the next, some species disappeared and some new species appeared.

Unable to break through the prevailing beliefs about life, Cuvier could not link these observations to the concept of evolution. Instead, staying within the boundaries of his beliefs, he explained his observations with the hypothesis of *catastrophism*:

1. A boundary between two layers of fossils represents a catastrophe in a region such as flood that had wiped out most of the species living in the region.

2. Later, some species moved from other places to the region where the catastrophe had occurred. This is how he explained the disappearance of some old species and appearance of some new species in fossils.

3. But there were too many catastrophes to account. Hence, Cuvier proposed that the geological events in the past used to occur suddenly and were driven by the mechanisms that were different from the mechanisms

operating today. This is known as the principle of *catastrophism*.

James Hutton (1726-1797): Some scientists, including a Scottish geologist James Hutton, disagreed with the hypothesis of catastrophism, and proposed that the geologic features of earth including fossils can be explained through mechanisms that operate gradually and are the same today as they were in the past. These mechanisms give rise to slow but continuous change, which accumulates over time into a profound change.

Charles Lyell (1797-1875): Charles Lyell, the foremost geologist of his time, incorporated Hutton's ideas into his principle of uniformitarianism: Mechanisms that cause change, themselves remain unchanged over time. For example, gravity today is the same gravity as it was at the beginning of time.

Jean-Baptiste de Lamarck (1744-1829): French biologist Jean-Baptiste de Lamarck was the first to publically defend the concept of evolution and to propose, in 1809, a testable mechanism for it. He proposed this mechanism in form of two principles:

- **Use and disuse:** The parts of the body that an organism uses become stronger and the parts that are not used deteriorate. For example, giraffes acquired their long necks by stretching their short necks to reach the leaves on high branches of trees.

- **Inheritance of characteristics:** Organisms can pass the modifications acquired through their lifetime to their offspring. For example, if giraffes can make their necks a little bit longer through their lifetime by stretching it, they can pass this length along to their offspring. If every generation does it, over a few generations, we will have giraffes with very long necks.

According to Lamarck, evolution is goal driven, and it occurs due to the inner drive of organisms to acquire more complexity. Lamarck's mechanism for evolution has been largely discredited through scientific evidence.

Alfred Russel Wallace (1823-1913): British naturalist, Alfred Wallace proposed his theory of evolution by natural selection along with Charles Darwin in 1858. Wallace and Darwin independently discovered the mechanism of natural selection in order to explain the data that they had accumulated.

Charles Darwin (1809-1882): British naturalist, Charles Darwin proposed his theory of evolution by natural selection along with Alfred Wallace in 1858. In the following year, Darwin published the book, *Origin of Species*, in which he explained the theory to a greater detail. Below is the abstract:

- The inherited traits in a population vary across individuals of the populations.

- The individuals of a population have the capacity to reproduce more offspring than the environment can support. This means that not all the offspring can survive. Then, who will survive?

- Remember, the inherited traits vary across the populations. So, the individuals with traits that are favorable in the given environment will survive and reproduce with greater rate than those individuals who do not have these traits. This process is called *natural selection*.

- As a result of natural selection operating over multiple generations, favorable traits will accumulate. This will continuously improve the fit between the population and its environment.

- The gradual accumulation of adaptations to a new environment may eventually give rise to a new species from an ancestral species.

Figure 1.5 illustrates these points.

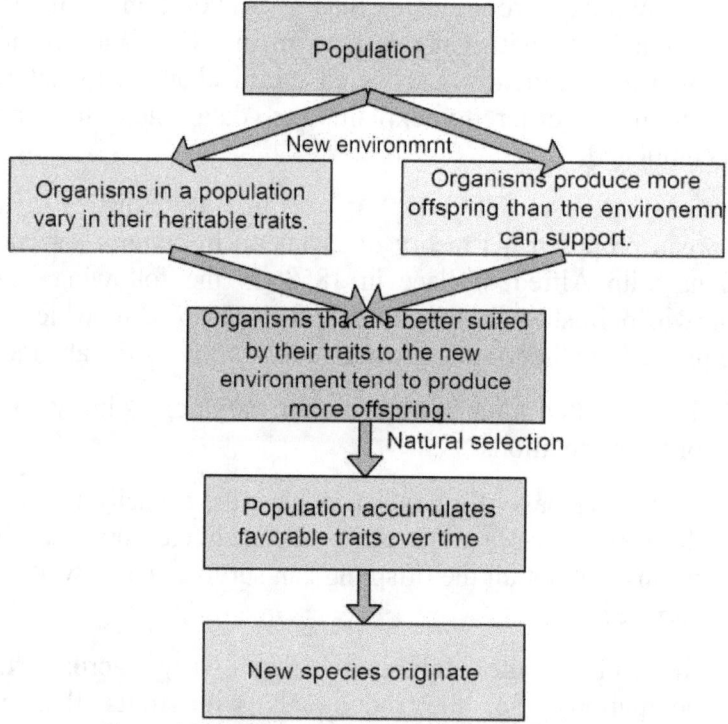

Figure 1.5 **Evolution through natural selection.**

The biggest problem Darwin struggled with was that he could not positively identify the source of variation among inheritable traits. Variations are the raw material for natural selection. Take them away, and there would be no evolution.

Gregory Mendel: The source of variations was revealed by Darwin's contemporary Gregory Mendel (1822-1884), an Austrian monk, in his publications based on the experiments that he performed on pea plants between 1856 and 1863. However, Darwin did not have access to Mendel's work, because Mendel died in obscurity in 1884, with his work (which would later bring a revolution in biology) ignored, unknown, or forgotten. Mendel is now known as the founding father of modern genetics. Within the framework of his work, the different versions of genes, called alleles, are the origin of variations of inherited traits among organisms.

Hugo de Vries (1848-1935): A Dutch botanist Hugo De Vries, along with other two scientists, rediscovered Mendel's work in 1900 and made the connection between it and the theory of evolution. Vries proposed that evolution occurs due to mutations in the Mendelian genes, which give rise to different versions of the genes called alleles. He published this theory of mutations in 1901.

Caution! Darwin was neither the first to introduce the concept of evolution, nor the first to introduce the mechanism for evolution. Lamarck was the first to propose a mechanism for evolution, which was later discredited through experimental evidence. Finally, both Wallace and Darwin independently proposed the principle of natural selection as mechanism for evolution, which has withstood the test of observations and experiments up to date.

Since the time of Darwin and Wallace, generations of scientists have worked to develop the theory of evolution to its modern form. You will see in the next chapter that other mechanisms, in addition to natural selection, that contribute to evolution have been discovered. However, natural selection remains at the core of the theory of evolution.

1.3 Salient Features of Natural Selection

Natural selection, once understood, seems the obvious mechanism to explain evolution. To understand it, one must know its main features:

1. Natural selection operates on heritable variations. Therefore heritable variations must exist for natural selection to be effective. In other words, evolution by natural selection cannot occur in a population in which all the organisms are genetically identical. There is no evolution without variation.

2. Natural selection is not random. It selects for the organisms with favorable traits and selects against the organisms that have unfavorable (harmful) traits.

3. Natural selection operates at the level of organisms. It selects organisms according to their phenotypes.

4. Natural selection is not goal-oriented.

5. Because natural selection does not create anything on its own, but operates on a given set of traits already present in the population, it does not lead to perfect organisms or species.

6. Natural selection improves the fit of the population within its environment.

7. Heritable traits that are favorable in one environment may be useless or even harmful in another environment.

> Caution! Evolution is not desire driven, it is not goal oriented, and it does not create a perfect product.

Driven by their beliefs and not scientific observations, some people still disagree with the concept of evolution. Their main argument is that it is just a theory.

1.4 Is Evolution Just a Theory?

One of the common attacks on the theory of evolution by its opponents is: it is *just a theory*. This one line attack can be countered with a one line defense: It is a not *just a theory*, it is a scientific theory. In science, a theory always means a scientific theory, and a scientific theory by its very definition is not *just a theory* or a speculation, but is a set of principles or explanations well supported by evidence and reproducible tests. Note the following three points:

- By definition, a scientific theory must be falsifiable, that is, lend itself to reproducible tests.

- In science, hypotheses and principles are organized into a theory only after a wide acceptance in the scientific community through extensive testing of their predictions and through a sound support by a large body of evidence.

- A scientific theory is continuously under scientific scrutiny and is subject to modification or rejection based on newly found facts or experimental results.

The theory of evolution is a well-tested scientific theory that has withstood extensive scientific scrutiny thus far. In the next section, we will explore some of the lines of evidence for the theory of evolution.

1.5 Lines of Evidence for Evolution

The power of a scientific theory is found in the predictions it makes. A scientific theory such as evolution can be tested by testing its predictions and by investigating if the theory explains the scientific observations and data. In this context, following are some of the lines of evidence for evolution:

Homologies: Because theory of evolution states that different species evolved from a common ancestral species, it predicts that we must be able to find similarities in characteristics among organisms of different species that resulted from a common ancestry. Such similarities are called *homologies*. This prediction is abundantly supported by observations. For example, there are similarities among basic underlying structures of the forelimbs of different animals such as arms of human, forelegs of elephants, flippers of penguins, and wings of flying birds. These homologies suggest that these forelimbs evolved from the same structure that existed in their common ancestor.

Analogies and Convergent Evolution: Similar environments can exist in different parts of the world. Thus the theory of evolution predicts that we must be able to find some similar characteristics among organisms from different species that do not share a common ancestor in the recent past, and therefore belong to different lineages of evolution. There similarities would have arisen from adaptations to similar environments. Such similarities are called analogies, and characteristics (traits) with these similarities are called *analogous characteristics*. This kind of evolution is called *convergent evolution*. Quite a few examples of analogous characteristics have been discovered. For instance, the wing surfaces of bats and insects are analogous structures. Evolutionarily, a bat is more closely related to human than to

14

insects. Even the very fact that bats, insects, and birds have wings despite being very different in their ancestry can be explained with the concept of analogous characteristics. They happened to fill the flying niche at the time and the locations in which the wings evolved.

Note. Sometimes the evidence from homological, analogical, and vestigial structures is presented under the title of *comparative anatomy*, which is the study of comparing the anatomy of different organisms and species.

More evidence for evolution emerges from the field of bio-geography.

Biogeography: Biogeography is the study of patterns in the geographical distribution of species and communities in the present and in the past. For example, the following biogeo-graphical patterns are consistent with the theory of evolution:

- An island's isolation from the rest of the world: It has been observed that many species living on islands are not found anywhere else in the world. Because islands are isolated places, the species that evolved there are not found anywhere else. This is consistent with the theory of evolution. The linear hierarchy of creationism with the doctrine of permanent species independent of each other cannot explain this observation. This is because if the species were created from scratch by the Creator permanently, they could have been anywhere, and the isolation pattern of some species does not follow from this doctrine.

- An island's connection with the nearest mainland: Most species living on an island resemble the species in a neighboring island or the species in the nearest mainland more than the species from a distant island or distant mainland with a similar environment. This is true even if the nearest mainland has a different environment than the island. It means that at some time, in the distant past, some species from the nearest mainland made it to the island and evolved into a new species in the new environment of the island.

- **Origin of the horse species:** The theory of evolution predicts that the oldest horse fossils should only be found in North America. This is because evolutionary trees constructed based on fossil record and anatomical data suggest that the modern horse species originated about five million years ago in North America, when it was not connected with South America. The fossil data collected so far has supported this prediction.

Paleontology: Paleontology is the study of fossils, which are the preserved remains or traces of organisms that lived in the past. The fossils on rocks are like the history of life written on rocks. The fossil record reveals a pattern of life on Earth over a very long time span, which is consistent with the predictions of the theory of evolution. Through fossils, present day species can be traced back and linked to older species which were different from the present species. Fossils also show that most of the species of the past have gone extinct. Fossil records also provide evidence for changes within a given group of organisms over time, and for the origin of new groups or species.

Vestigial Structures: If multiple species evolved from a common ancestor, then we must be able to find in the existing species some useless remnants of structures that were useful for the ancestors. Such structures have been found and are called vestigial structures. For example, we human still have coccyx, the tail bone, which is the remnant of the lost tail. Another example of our vestigial structure is

16

goose bumps, which had very useful function in our ancestral species: to raise the body hair, making the organism appear larger in order to scare off the predators, but are not useful in modern day.

Comparative Embryology: Comparative embryology is the study that compares the development of embryos of different species. Embryology provides evidence for evolution by revealing the striking similarities among the embryos of very different species. Some species have almost identical embryos in the beginning of development. For example, the embryos of a bird, fish, human, and rabbit look very similar in the early stages of development, as all of them have gill slits, a tail, and a two-chambered heart. As the development progresses, these embryos take different paths and the resemblance decreases. This suggests that the development mechanism started with a common ancestor, and was modified for different species as they evolved in different environments.

Field Studies: There has been a continuous stream of evidence for evolution from thousands of field studies by biologists. One of the clearest examples is the study of drug-resistant bacteria. It has been observed in the field of medicine and proven in the laboratory that pathogenic bacterial populations evolve drug resistance through natural selection. In the beginning, there are a very few individuals in the population that have drug resistance capability, so the antibiotic drug works. However, the repeated use of the drug kills most of the drug-vulnerable individuals of the population, so the drug resistant individuals multiply over generations. Therefore, the drug becomes ineffective, as the bacteria have evolved to survive it.

> Note. Evolution through natural selection consistently improves the fit between the population and its environment.

Molecular Biology and Biochemistry Including Genetics: Because evolution is based on inherited traits and because traits are inherited through genes, you can imagine that genetics, the study of genes, would have a great say in crediting or discrediting the theory of evolution. Indeed, genetics have produced a formidable body of evidence for evolution. All organisms use the same genetic code to reproduce and main-tain life. This means that the genetic code is universal. This remarkable fact strongly suggests that all diverse groups of organisms have descended from one common ancestral group that used this genetic code. Further evidence states that organisms from different species also share genes, which can be owed to common ancestors as well. These similarities are also called *molecular homologies*.

> Caution! Heritable variations across the organisms of a population are the raw material for natural selection; without them natural selection has nothing to operate on.

1.6 More Definitions of Terms and Concepts

In this section, we define important terms and concepts related to the topic at hand that are not already discussed in this chapter or we compare those concepts that have been discussed but not adequately compared with one another closely.

Catastrophism, Uniformitarianism, and Gradualism: mind the gap. Note how these three terms compare with another:

- The doctrine of *catastrophism* states that events in the past occurred suddenly and were caused by mechanisms that were different from the mechanisms operating today.

- The principle of *uniformitarianism* states that the mechanisms that bring about change, themselves remain unchanged over time.

- *Gradualism* means the changes occurs slowly and can accumulate over time into a big profound change.

Extinct Versus Extant: The extinct species are those that existed in the past, vanished, and do not exist today; whereas extant species are those that still exist today.

Fascinating Fact! Based on evidence, scientists estimate that 99 percent of all species that ever existed on our planet are already extinct.

Homologous versus Analogous Traits: Homologous traits share common ancestry but not necessarily similar functions, whereas analogous traits share similar functions but not common ancestry. For example, all vertebrates such as birds, fish, frog and human have skeleton, which is a homologous trait because the vertebrates inherited it from their common ancestor. The bats, birds, and insects share wings, which in this context are called analogous trait because they evolved in each of these groups independently to serve the same function.

Evolutionary Pressure: Sometimes also called selective evolutionary pressure or selective pressure, *evolutionary* *pressure* is any environmental condition that favors (or disfavors) the individuals of a population selectively (unequally). In general, it is a factor or a condition that contributes to triggering one of the evolutionary forces (or mechanisms) such as natural selection to begin operating on the population. For example, large seeds put a selective evolutionary pressure on a population of birds because it would favor birds with large beaks against the birds with small beaks.

Differential Reproductive Success: This refers to the variation in success that organisms in a population with different traits have in reproducing viable and fertile offspring. Differential reproductive success, a necessary result of natural selection, occurs because organisms in the population vary in their ability to survive and reproduce.

Artificial Evolution and Artificial Selection: Artificial evolution is the evolution of a species by using non-natural means. It has been practiced by humans for thousands of years by selectively breeding domesticated plants and animals to increase the desirable traits. Such a selection is called *artificial selection* as opposed to natural selection in which the natural environment makes the selection. For example, broccoli, Brussels sprouts, cabbage, cauliflower, kale, and kohlrabi are all artificially evolved through artificial selection from wild mustard. In his book, *The Origin of Species*, Darwin mentioned artificial selection to craft his argument for evolution.

Caution! Artificial evolution is goal-oriented, whereas natural evolution is not.

1.7 In a Nutshell

The theory of evolution, originally introduced by Darwin and Wallace, is based on the concept of *descent with modifications from a common ancestor*, which explains the three widely observed characteristics of life: unity, diversity, and adaptations. Natural selection, the mechanism of evolution presented by Darwin and Wallace, explains how evolution occurs and how evolutionary change makes a population fit better with its environment, but they did not explain the cause of the trait variations responsible for evolution.

Since then the theory of evolution has come a long way. By now, through the work of many scientists, we know not only the cause of variations but also that evolution occurs at much smaller scale, called microevolution, which accumulates to the evolution that Darwin and Wallace observed, at macro scale and is called macroevolution. Furthermore, scientists have also discovered other mechanisms of evolution in addition to natural selection. We will explore these topics in the next chapter.

1.8 Review Questions

1. Which of the following statement about Lamarck's hypothesis is false?

   A. Organisms strive to become more complex.

   B. Use and disuse determine the change in traits of organisms, which they pass down to the next generation.

   C. Evolution is a slow process that happens over generations through inheritance of characteristics acquired by organisms in their life time.

   D. Lamarck's hypothesis is not a scientific hypothesis.

2. An adaptation is a feature that:

   A. A population develops in order to survive in a certain environment.

   B. An organism develops in order to survive in a certain environment.

   C. A population uses to survive through a certain environment.

   D. A change that is produced by a certain environment in the organisms of a population.

3. *Descent with modification* explains:

   A. Adaptation

   B. Unity underlying all organisms

   C. Diversity among all organisms

   D. All of the above

   E. A and B

4. Which of the following assumptions was not part of the theory of evolution introduced by Charles Darwin?

    A. Populations often produce more offspring than the resources in their environment can support.

    B. Traits are passed to offspring through discrete particles called genes.

    C. Natural selection is not random.

    D. Inherited traits vary across the populations.

5. Which of the following is the best example of evolution in humans?

    A. Skin color of a population living in a hot region became darker over generations.

    B. When you move from high light area to a low light area in your house, you cannot see much, but slowly your eyes adjust, and you can see much better.

    C. Your skin becomes tanned after multiple exposures to the sun at the beach.

    D. You can drink much more after eating a salty food.

    E. All of the above

    F. None of the above

6. A strain of pathogenic bacteria causes a disease. An antibiotic drug was manufactured to treat the patients. At first the drug was very effective. However, gradually the drug became less effective. What is the reason?

    A. The bacteria developed capabilities in order to adapt to the environment of antibiotics.

B. The antibiotic caused a mutation in the bacteria that gave rise to new genes in the bacteria, which are resistant to the antibiotics.

C. Even before the antibiotics were used, there were a few bacteria in the population that had drug-resistant genes, and they multiplied as the bacteria without these genes were killed by the antibiotics.

D. All of the above

E. None of the above

7. The arm of a human and a flipper of a penguin (used for swimming) appear somewhat different and both have different functions. Nevertheless, the underlying anatomy of these two structures is basically the same. This makes these structures as an example of:

A. Analogous structures

B. Convergent evolution

C. A and B

D. Homologous structures

E. Geographic isolation

8. The bat wings and bird wings were evolved independently in separate lineages. The flight surfaces of a bat wing and a bird wing are examples of:

A. Analogous structures

B. Convergent evolution

C. A and B

D. Homologous structures

E. Geographic isolation

9. Natural selection would not work if:

    A. The environment does not change.

    B. There are no genetic variations among a population.

    C. There is only one population in a community.

    D. There are only two populations in the community and both populations are equally strong, that is, physically.

10. Evolution is:

    A. Heritable change as a result of descent with modification

    B. Natural selection

    C. Change as a result of use and disuse

    D. As a result of genetic variations

    E. A and D

1.9 Answer Key

1. D	6. C
2. C	7. D
3. D	8. C
4. B	9. B
5. A	10. E

Notes:

Q1. Lamarck's hypothesis is scientific because it is falsifiable and testable.

Q2. Environment does not create the feature that helps an organism to survive; it only selects the feature that already exists and multiplies among the population over generations.

Q4. Darwin did not know about genes.

Q10. Natural selection is one of several mechanisms through which evolution can occur.

Chapter 2

Microevolution and Mechanisms of Evolution

2.1 Microevolution and Macroevolution: The Big Picture

Microevolution is the evolution of the gene pool of a population in terms of changes in allele frequencies. As you learned in Chapter 1, trait variations are necessary for evolution to occur. Because phenotypes (visible traits) arise from genotypes (genetic traits), therefore the trait variations of an organism's phenotype that are necessary for evolution to occur, arise from the organism's genotype variations. Genotype variations are due to multiple alleles, as shown in Figure 2.1. This multitude of alleles is called the gene pool of a population. It is the sum total of all the alleles (versions) of all the genes in all the organisms of a population.

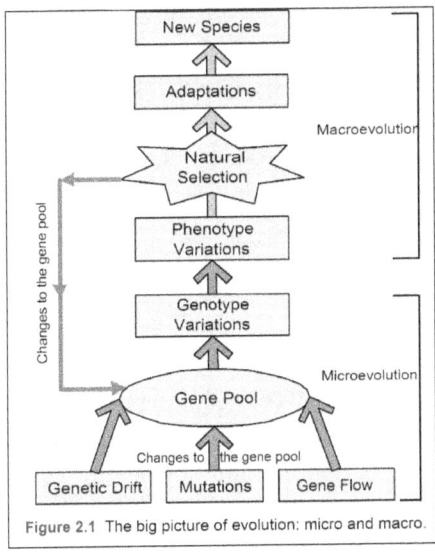

Figure 2.1 The big picture of evolution: micro and macro.

Biology of Evolution and Systematics, by Paul Sanghera
Copyright © 2015 Infonential.

Caution! Although it is true that inherited phenotypes arise from genotypes, environmental influences also play their role in shaping a phenotype. In general, a phenotypes arises as a combined effect of inherited genotype and environmental influences:

Genotype + Environment ➡️ Phenotype

However, the genetically determined part of a phenotype is the only part that is passed down to the offspring.

The population evolves as the frequencies of different alleles in the population gene pool change. These changes in frequencies can occur over multiple generations or simply from one generation to the next. Microevolution is micro (small) in two respects. First, it is happening at the molecular (micro) level of genes, and second, we can measure it from one generation to the next, even when it may not be visible at the organism or species (macro) level. The changes accumulated from microevolution over multiple generations may give rise to macroevolution, the evolution that happens at the level of species, giving rise to new species or groups of organisms. This macroevolution is the evolution that Darwin and Wallace observed.

Evolution can be explained by using a number of different mechanisms.

2.2 Mechanisms of Evolution

There are five mechanisms that can bring about change in allele frequencies, genotype frequencies, or both: mutations, sexual reproduction, natural selection, genetic drift, and gene flow.

Mutation, as it was explained in Chapter 1, is a change in the nucleotide sequence of DNA that may arise for various reasons such as errors in the DNA replication. Mutation is the fundamental (or primary) cause of (new) alleles. Mutations occur slowly over time. For example, for us humans the rate of mutation is about 175 per person per generation.

The process of sexual reproduction, although responsible for genotype variations, just shuffles around the existing alleles, and therefore does not change the allele frequencies. However, this is true only under the assumption of random mating, as non-random mating can change genotype frequencies. So, most of the near term changes in the gene pool result from genetic drift, gene flow, and natural selection.

Caution! Sexual reproduction that happens through random mating within a population does not change the allele frequencies nor the genotype frequencies of the population. However, non-random mating can change the genotype frequencies of the population. Even though non-random mating does not change the allele frequencies on its own, it can change allele frequencies dramatically when combined with natural selection.

As introduced in Chapter 1, natural selection operates on the phenotypes or organisms. It also indirectly changes the gene pool by selecting for a certain section of the population whose allele frequencies are different from the overall allele frequencies of the population.

Gene flow means the transfer of genes or alleles into or out of a population. This can happen for example as a result of migration of individuals into a population that have the ability to reproduce viable and fertile offspring with the organisms already existing in the population. Gene flow can add beneficial, neutral, or harmful alleles into the population.

Genetic drift is the process in which the events based on chance or probability cause changes in allele frequencies of the population's gene pool from one generation to the next. Genetic drift has noticeable effect only in small populations. For example, in a population of one hundred people, if there are only ten blue-eyed individuals and eight of them are killed in an accident before getting married and therefore do not produce offspring; this chance occurrence will cause a huge change in the blue-eye allele frequency of the population. If the accident happened in a population of thousands where there were many blue-eyed individuals, then the effect of the genetic drift would have gone unnoticed.

Problem 2.1 Allele Frequency

Consider a population of a bird species with a total of 1000 birds. The wing color is determined by two alleles corresponding to a pigment gene: the dominant allele B and the recessive allele b. In this population, 400 birds have dark-blue wings due to genotype BB, 250 have medium-blue wings due to genotype Bb, and 350 have white wings due to genotype bb. Calculate the frequency of alleles B and b in the population.

Solution:

Total number of alleles = $1000 \times 2 = 2000$

Number of B alleles = $400 \times 2 + 250 = 1050$

Frequency of B = $p = 1050/2000 = 0.525 = 52.5\%$

Number of b alleles = $350 \times 2 + 250 = 950$

Frequency of b = $q = 950/2000 = 0.475 = 47.5\%$

The two most significant types of genetic drift are the bottleneck effect and the founder effect.

The Bottleneck Effect: This genetic drift occurs as a result of a sudden event in the environment that reduces the population size, resulting in a population after the event that has a different gene pool than the population before the event. Natural disasters such as fire or flood or ruthless human interaction with nature can cause such events.

The Founder Effect: This genetic drift occurs when a few organisms depart from a large population and make their own population in a new environment. The allele frequencies in the gene pool of this new population will be different from those in the gene pool of the original population. An example of the founder effect could be that a few members of a population on the mainland are flown away by a powerful storm to an island.

Note. Because genetic drift works more effectively on small populations one of its possible effects is the fixation of alleles resulting in decreased genetic variation. An allele is said to be fixed when it is the only allele corresponding to its gene in the population gene pool, that is, it has a frequency of 1.

Natural selection, the only mechanism that consistently causes adaptive evolution, needs to be explored further.

2.3 Three Types of Natural Selection

Based on the environment, natural selection in a given population operates in one of the three modes, each mode resulting in different outcome. These three modes are: stabilizing selection, directional selection, and disruptive selection. As illustrated in Figure 2.2, stabilizing selection favors intermediate variants, i.e. medium size beaks, directional selection favors variants leaning toward one of two extremes, i.e. short beaks; and disruptive selection favors variants at both extremes, i.e. short and long beaks.

31

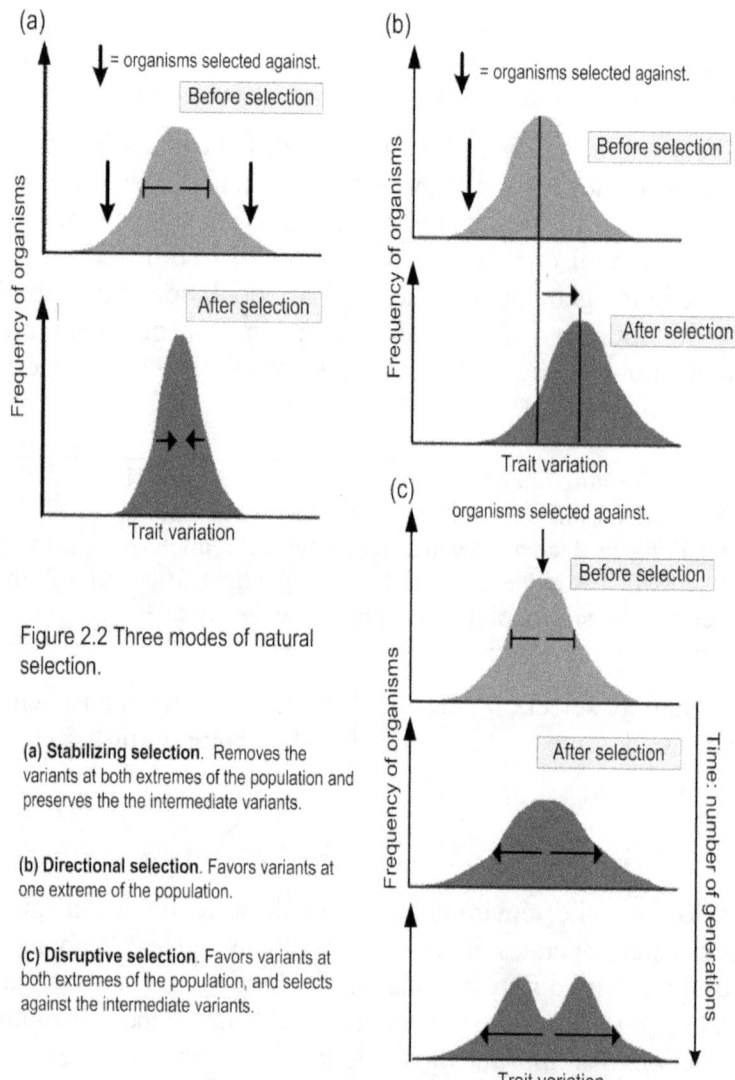

Figure 2.2 Three modes of natural selection.

(a) **Stabilizing selection**. Removes the variants at both extremes of the population and preserves the the intermediate variants.

(b) **Directional selection**. Favors variants at one extreme of the population.

(c) **Disruptive selection**. Favors variants at both extremes of the population, and selects against the intermediate variants.

Directional selection, disruptive selection, or stabilizing selection may happen when the environment of a population changes or the population moves into a new environment.

For example, if a bird population moves from an environment where all sizes of seeds are equally available to an environment where more small seeds are available than large and intermediate seeds then this change will favor the short beaked birds, and directional selection will result. Note that stabilizing selection and directional selection reduce variants (diversity) of the population. Stabilizing selection, as it reduces variants, tends to maintain the status quo for a specific variant.

The common characteristic of all three types of natural selection is that the variants that help the organism to enhance survival and reproduction are favored. Therefore, the mechanisms of natural selection consistently causes adaptive evolution, that is improves the fit between the population and its environment. Other mechanisms such as genetic drift and gene flow may or may not cause adaptive evolution.

Caution! Natural selection is not the only mechanism of evolution, but it is the only mechanism that consistently improves the fit between the population and its environment. In other words, it consistently leads to adaptive evolution.

Microevolution can be measured quantitatively by using the Hardy-Weinberg principle.

2.4 Hardy Weinberg Equilibrium: Measuring Evolution

Allele frequencies represent the relative proportions of different alleles in a population. The Hardy-Weinberg principle states that, if a population is not evolving, then the frequencies of alleles and genotypes in that population will remain constant from one generation to the next. Further, this principle allows us to predict what the genotype frequencies will be in a non-evolving population. By looking at how allele frequencies change over time from generation to generation, we can see if a population is evolving, as well as how much it's evolving.

We can conclude that a population is evolving if its genotype frequencies differ from those predicted by the Hardy-Weinberg principle or theorem.

Hardy-Weinberg Theorem: If only Mendelian segregation and recombination of alleles are at work in a population, the allele frequencies and the genotype frequencies of the population will remain constant from generation to generation, and therefore the evolution will not occur.

This principle can be expressed in the following mathematical equation:

$$p^2 + q^2 + 2pq = 1 \qquad (2.1)$$

Where p is the frequency of the dominant allele in the population gene pool for a trait (genotype) and q is the frequency for the recessive allele for the same genotype. Such a non-evolving population is said to be in Hardy-Weinberg equilibrium. Equation 2.1 is derived in the box. These frequencies are called expected frequencies. If the measured frequencies agree with the expected frequencies, the population is in Hardy-Weinberg equilibrium, that is, not evolving.

Derivation of the Hardy-Weinberg Theorem

Consider a genotype (representing a trait) with dominant allele A and recessive allele a. Assume that the allele A has the frequency p in the population gene pool and the allele a has the frequency q:

$$p + q = 1 \quad (2.2)$$

By taking the square of both sides of equation 2.2, we obtain:

$$p^2 + q^2 + 2pq = 1 \quad (2.1)$$

Equation 2.1 is the mathematical form of the Hardy-Weinberg theorem. It is a quick mathematical derivation. However, to realize what the terms in the left hand side of the equation mean, let's derive the same equation through a more physical logic.

The total probability that an organism produced from this trait through random mating will be homozygous for the dominant allele, or homozygous for the recessive allele, or heterozygous is equal to one, because one of these three cases will be true. The probability that one parent will pass down the dominant allele to the child is p. The probability that both parents will pass down the dominant allele is, by the product rule of probabilities, equal to p×p $= p^2$.

Therefore, the probability for an organism to be homozygous in dominant allele, which is the same as the frequency of organisms in the population with the dominant allele, is p^2. Similarly, we can argue that the probability for an organism to be homozygous in recessive allele, which is the same as the frequency of organisms in the population with the recessive allele, is q^2.

Now, the probability that a dominant allele from a male parent will meet with the recessive allele of the female parent is, again by the product rule of probabilities, equal to pq. Similarly, the probability that a recessive allele from the male parent will meet with the dominant allele of the female parent is also pq. Therefore, the probability for an organism to be heterozygous, which is the same as the frequency of organisms in the population that are heterozygous for the trait, is 2pq.

Because an organism could either be heterozygous, or homozygous for the dominant allele, or homozygous for the recessive allele; the sum of these three probabilities must be 1. Therefore:

$$p^2 + q^2 + 2pq = 1 \quad (2.1)$$

Now, through this derivation, it is clear that p^2 is the frequency of organisms in the population that are homozygous for the dominant trait, q^2 is the frequency of organisms that are homozygous for the recessive trait, and 2pq is the frequency of organisms that are heterozygous.

Biology of Evolution and Systematics

Remember, for a population to be in Hardy-Weinberg equilibrium, it has to meet the following requirements:

1. **Absolute randomness in mating:** This means all members of the population breed and choose their breeding partners absolutely at random, and everyone produces the same number of offspring. Under these conditions there is no change in allele frequencies of the gene pool due to sexual reproduction. In reality these conditions are never met and therefore evolution from sexual selection does occur.

2. **No mutation:** Mutations modify the gene pool by actions such as altering alleles, deleting genes, and duplicating genes. No mutation means no changes (in allele frequencies) from the mutations to the gene pool. There is plenty of evidence in the history of life on Earth to show that mutations do happen.

3. **No gene flow:** This condition means that no genes are flowing out of or into the population, which means no genetically diverse individuals are immigrating into or emigrating out of the population. In reality this condition is not commonly met, hence evolution from gene flow does take place.

4. **No genetic drift:** This situation is realistic only in extremely large populations. The smaller the population, the greater the probability for allele frequencies to be changed by chance events, resulting in the evolution of the population.

5. **No natural selection:** Natural selection does not affect the gene pool directly. It operates on individual organisms, and if an individual is selected out, all of its genes are expelled out of the gene pool of the population. But because natural selection is not random, that is, it favors some genes (corresponding to favorable traits)

against others, it changes the allele frequencies of the gene pool by passing alleles to the next generation in different proportions than they are in the current generation. Therefore, for a population not to evolve, natural selection must not be operating. If natural selection is operating, and therefore the gene pool is changing, then the population is evolving.

Note. In order for a population to be in the Hardy-Weinberg equilibrium it cannot evolve. Only Mendelian segregation and recombination of alleles are allowed. No other process that may change the genetic makeup of the population is allowed. You can imagine that it is impossible to maintain all the five conditions of Hardy-Weinberg equilibrium in nature. Hence, Hardy-Weinberg principle is a very good means to realize that evolution does happen.

Various concepts in microevolution and Hardy-Weinberg theorem are illustrated in problem 2.2.

Problem 2.2: Microevolution and the Hardy-Weinberg Theorem

A scientist is studying the evolution of a population of insects. The dominant pigment allele is G for green and the recessive pigment allele is g for yellow. The wings of insects that are homozygous for the dominant pigment gene (GG) or are heterozygous (Gg) are therefore green, and the insects that are homozygous for the recessive gene (gg) are yellow in color. At a given point in time, the scientist took the following measurements:

Biology of Evolution and Systematics

Table 2.1 Measurements

Observed Color	Genotype	Observed Number of Organisms
Green	GG	50
Green	Gg	240
Yellow	Gg	150

Q1. What are the observed frequencies for the G and g alleles?

Answer:

Total number of organisms = 50 + 240+150 = 440

Total number of pigment alleles = 2 × 440 = 880

Number of G alleles = 2 × 50 + 1 × 240 = 340

Frequency of G = p= 340/880 = 0.386 = **38.6 %**

Number of g alleles = 2 × 150 + 1 × 240 = 540

Frequency of g = q= 540/880 = 0.614 = **61.4 %**

Q2. What would be the expected number of individuals of different genotypes, if no evolution occurred?

Answer:

GG = Expected frequency of GG individuals times the number of total individuals = p^2 × 440 = 0.386^2 × 440 = 65

Gg = Total number of heterozygous individuals = 2 pq × 440= 2 × 0.386 × 0.614 × 440 = 209

gg = Total number of individuals with gg genotype = q^2 × 440 = 0.614^2 × 440 = 166

Let's put it in a table:

Table 2.2 Comparison between measurements and predictions of the Hardy Weinberg equilibrium.

Color	Genotype	Observed Number of Organisms	Expected Number of Organisms
Green	GG	50	65
Green	Gg	240	209
Yellow	Gg	150	166

Q3. Look at Table 2.2 above and determine if the population is in Hardy-Weinberg equilibrium. If not, specify what's happening?

Answer:

Because the expected numbers from the Hardy-Weinberg equilibrium do not match with the measured numbers, the population is not in Hardy Weinberg equilibrium. Therefore the population is evolving. The evolution is favoring heterozygotes (Gg) against homozygotes (GG and gg).

Q4. Calculate the relative fitness of all the three genotypes: GG, Gg, and gg.

Answer:

Relative fitness can be calculated by the proportions of observed numbers divided by the expected numbers:

GG:Gg:gg = 50/65:240/209:150/166=

0.769:1.15:0.904=

0.769/1.15:1.15/1.15:0.904/1.15=

0.669:1:0.786

Although the Hardy-Weinberg equation (Equation 2.1) is written for genes with only two alleles, the Hardy-Weinberg

39

equilibrium can be easily extended to genes with more than two alleles. Problem 2.3 illustrates how.

Problem 2.3. In humans, the blood type is determined by a single gene with three alleles: I^A, I^B, and i. Consider a human population is in equilibrium and not evolving. If the frequency of the I^A allele is 0.3 and the frequency of the I^B allele is 0.2, calculate the percentage of population with different blood types:

Blood type A: $I^A I^A$ or I^A I

Blood type B: $I^B I^B$ or I^B i

Blood type AB: $I^A I^B$

Blood type O: ii

Solution:

Frequency of I^A = 0.3 (given)

Frequency of I^B = 0.2 (given)

Frequency of i = 1-0.3-0.2 = 0.5

Frequency of $I^A I^A$ = 0.3×0.3 = 0.09

Frequency I^A i = 2× 0.3×0.5 = 0.3

Population with blood type A = 0.09 + 0.30 = 0.39 = **39%**

Frequency of $I^B I^B$ = 0.2×0.2 = 0.04

Frequency of I^B i = 2×0.2×0.5 = 0.20

Population with blood type B = 0.04 + 0.20 = 0.24 = **24%**

Population with Blood type AB = Frequency of $I^A I^B$ = 2×0.2×0.3 = 0.12 = **12%**

Population with Blood type O = Frequency of ii = 0.5×0.5 = 0.25 = **25%**

Mutation being the primary cause of genetic diversity contributes to evolution in general. However, mutations do not occur frequently. Therefore, genetic drift, gene flow, and natural selection are the major mechanisms of microevolution, which leads to macroevolution. Natural selection contributes to microevolution indirectly by operating on phenotypes of organisms. In Section 2.3, we discussed how natural selection creates trait variations in three modes: direction selection, disruptive selection, and stabilizing selection. Natural selection also has other function-specific names that we will discuss in the next section along with some other concepts.

2.5 More Definitions of Terms and Concepts

In this section, we define important terms and concepts related to the topic that are not already discussed in this chapter.

Relative Fitness: The relative fitness of an organism in a population is the contribution that the organism makes to the gene pool of the next generation as compared to the contributions of other organisms in the population. This is a measure of the reproductive success of the individual compared to other individuals. See question 4 of Problem 2.2 for further clarification.

Sexual Selection: This is a type of natural selection that operates on a set of traits directly involved in obtaining mates. In other words, organisms with certain inherited traits have an advantage in obtaining mates over those organisms that lack these traits. There are two main forms of sexual selection: intrasexual selection and intersexual selection.

Biology of Evolution and Systematics

- **Intrasexual Selection:** This is the sexual selection in which an organism exerts selective pressure on other organisms of the same sex. An example is a male preventing other males from accessing a group of females.

- **Intersexual Selection:** This is the sexual selection in which an organism exerts selective pressure on other organisms of the opposite sex, for instance, by being selective in choosing a mate. An example is a female choosing a male mate with specific traits. Sexual selection gives rise to sexual dimorphism.

- **Sexual Dimorphism:** This refers to the difference between the males and females in secondary sexual traits such as behavior, color, ornamentation, and size.

Diploidy: Diploidy is the characteristics of organisms whose cells have two sets (2n) of chromosomes, one set inherited from each parent. Diploidy contributes to genetic variation; recessive alleles that are less favorable or harmful in a given environment can escape the axe of natural selection by propagating through heterozygotes.

Balancing selection: This form of natural selection maintains multiple forms of a phenotype in a population. It is called balanced polymorphism. Two examples of balancing selection are frequency dependent selection and heterozygote advantage.

Frequency dependent selection: This is the natural selection in which the relative fitness of a phenotype depends on the relative number (frequency) of the phenotype in the population, that is, how common the phenotype is. It maintains variation in an oscillatory or cyclic mode: increases the frequency of less common form of phenotype and decreases the frequency of more common phenotype; when the more common phenotype becomes less common, it's favored again, and the cycle continues preserving both phenotypes.

Heterozygote advantage: The heterozygote advantage, as the term suggests, is the natural selection in which heterozygous organisms in a population has greater relative fitness than the homozygous organisms. Because heterozygotes have multiple alleles, heterozygote advantage tends to maintain variations in the population gene pool.

Inbreeding: This is a type of nonrandom mating in which members of a population only mate with their close relatives. The effect is change in genotype frequencies in a population.

> **Caution!** Natural selection such as directional and stabilizing selection reduces variation by eliminating some phenotypes. This effect is counterbalanced by mechanisms such as diploidy, balancing selection, heterozygote advantage, and frequency dependent selection, which help preserve genetic variations in a population.

2.6 In a Nutshell

Evolution is caused by variations in inherited traits, which in turn are caused by genetic variations whose primary origin is mutation. Macroevolution, as observed by Darwin and Wallace, is the accumulation of microevolution, which is measured as the change in the gene pool of a population in terms of change in allele frequency. The major mechanisms of microevolution are genetic drift, gene flow, and natural selection. Natural selection, although it operates on phenotypes at the level of organisms, contributes to microevolution indirectly because phenotypes are associated with genotypes.

How does microevolution accumulate to give rise to macroevolution events such as the rise of a new species, called speciation? In other words, how do species really originate or evolve? This is the topic of the next chapter.

2.7 Review Questions

1. Assume that *like with like* mating is happening in a population. For example, organisms with brown skin mate with organisms also with brown skin, organisms of short height mate with organisms also of short height, and so on. Which of the following is not a likely effect of this non-random mating?

   A. Change in allele frequencies.

   B. Decreases in heterozygotes.

   C. Increase in homozygotes.

   D. Possible exposure of more phenotypes to natural selection.

2. Which of the following statements correctly describes relative fitness of individuals in a population?

   A. Organisms with low relative fitness gets less mates than other organisms.

   B. Organisms with high relative fitness have higher success in producing viable fertile offspring.

   C. Organisms with high relative fitness gets more food than do other organisms in the population.

   D. Organisms with high relative fitness are physically stronger than other organisms.

3. A scientist has concluded from data that allele frequencies in the gene pool of a population have changed randomly by chance. If this is true, what mechanism of evolution is operating in this population?

   A. Natural selection

   B. Gene flow

   C. Genetic drift

   D. Sexual selection

4. Which of the following statements correctly reflects the role of chance in evolution?

   A. Natural selection operates in a random fashion.

   B. New alleles are produced by chance by mutation.

   C. The basis of macro evolution is chance because it is an accumulation of micro changes that occur by chance events such as mutation and genetic drift.

   D. Chance events help increase the frequency of an allele, which in turn increases the reproductive fitness of a population.

5. In this information age of frequent communication, travel, and migration, the world has really become a global village. This will likely make which of the following mechanism less effective in the evolution of humans?

   A. Natural selection

   B. Gene flow

   C. Mutation

   D. Genetic drift

   E. Random mating

Biology of Evolution and Systematics

6. Assume that non-random mating is very common in a population. This situation will lead to which of the following?

   A. Allele frequencies of the population will change.

   B. Genotype frequencies of the population will change.

   C. Allele frequencies will change if natural selection is operating effectively.

   D. All of the above are true.

   E. Both B and C are true.

7. **True or False:** If there were no genetic variations, there would be no evolution.

8. A population is in Hardy-Weinberg equilibrium. The frequency of a dominant allele corresponding to a specific gene (with two alleles) in the gene pool is 0.3. What percentage of the population is heterozygous for this gene?

   A. 9%

   B. 49%

   C. 6%

   D. 21%

   E. 42%

9. A population is in Hardy-Weinberg equilibrium. If the frequency of the homozygous-dominant genotype (AA) is 0.36, what percentage of the population is heterozygous (Aa)?

   A. 64%

   B. 48%

C. 16%

D. 8%

10. Which of the following consistently results in adaptive evolution?

   A. Stabilizing selection

   B. Natural selection

   C. Directional selection

   D. Disruptive selection

   E. All of the above

2.8 Answer Key

1. A	6. B
2. B	7. True
3. C	8. E
4. B	9. B
5. D	10. E

Notes:

Q1. Allele frequencies do not change as a result of non-random mating, only genotypes (and hence phenotypes) do.

Q5. Ease of communication and access to anywhere in the world greatly reduces the probability of a small human population living in isolation for generations.

Q6. Non-random mating on its own only changes genotype frequencies and not allele frequencies. However, random mating combined with natural selection can make significant changes in allele frequencies.

Q8.

Frequency of dominant allele = p = 0.3

Frequency of recessive allele = 1-0.3 = 0.7

Frequency of heterozygotes = 2pq = 2×0.3×0.7 = 0.42 = 42%

Q9.

Frequency of the homozygous-dominant genotype (AA) = p^2 = 0.36

Frequency of dominant allele A = p = sqrt (0.36) = 0.60

Frequency of recessive allele a = q=1-0.60 = 0.40

Frequency of heterozygotes (Aa)= 2pq = 2×0.60×0.40 = 0.48 = 48%

Q10. Stabilizing, directional, and disruptive selections are just different modes of natural selection; and natural selection consistently leads to adaptive evolution.

Chapter 3

Evolution of Species

3.1 Evolution of Species: The Big Picture

A big mystery during Darwin's time, that Darwin referred to as *mystery of mysteries,* was speciation, the rise of new species. Darwin explained speciation in terms of *descent with modification.* He defined it as the process in which one species split into two or more species due to the modifications that evolved over generations.

Speciation is no longer a big mystery thanks to the progress made in molecular biology which includes genetics and biochemistry. Now we know that the big picture of speciation, as presented in Figure 3.1, involves many things from the gene pool to speciation and beyond. According to the image,

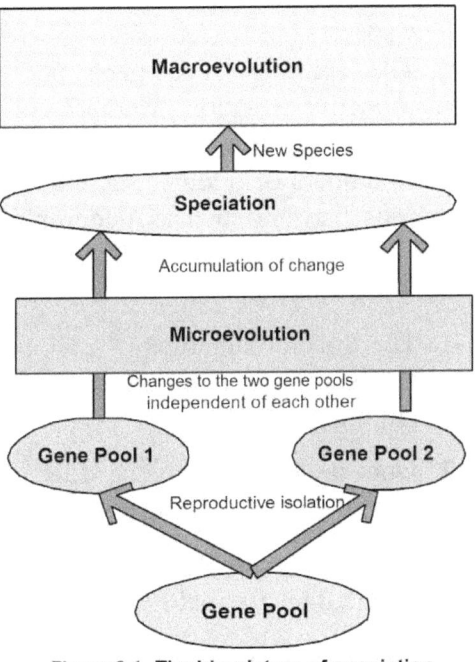

Figure 3.1 **The big picture of speciation**

Biology of Evolution and Systematics, by Paul Sanghera
Copyright © 2015 Infonential.

speciation is the link between microevolution and macroevolution. It is the link that is visible in the macroworld. In other words, macroevolution begins with speciation.

The biological definition of species is a group of organisms that have the potential to interbreed and produce viable and fertile offspring, and do not have the same potential to breed with organisms of other groups. A species may be geographically distributed into multiple subgroups called populations. Gene flow within a population (or a set of populations) of a species binds the population (or a set of populations) together into one species, whereas reproductive isolation provides the necessary ground for the formation of a new species.

The splitting of a gene pool occurs as a result of reproductive isolation. Members of one section of a species becomes unable to produce viable and fertile offspring with the members of another section of the same species due to various barriers such as geographic or habitat isolation. In this case, as illustrated in Figure 3.2, speciation can occur in the following steps:

1. The three populations (P_1, P_2, and P_3) are bound together as a species due to the gene flow (represented by arrows) among them.

2. Due to a barrier, gene flow between P_1 and P_2 and between P_1 and P_3 stops.

3. The population P_1 begins to genetically diverge from the populations P_2, and P_3.

4. Population P_1 has a gene pool which is different from the gene pool of populations P_2 and P_3.

5. Population P_1 over generations has given rise to a new species.

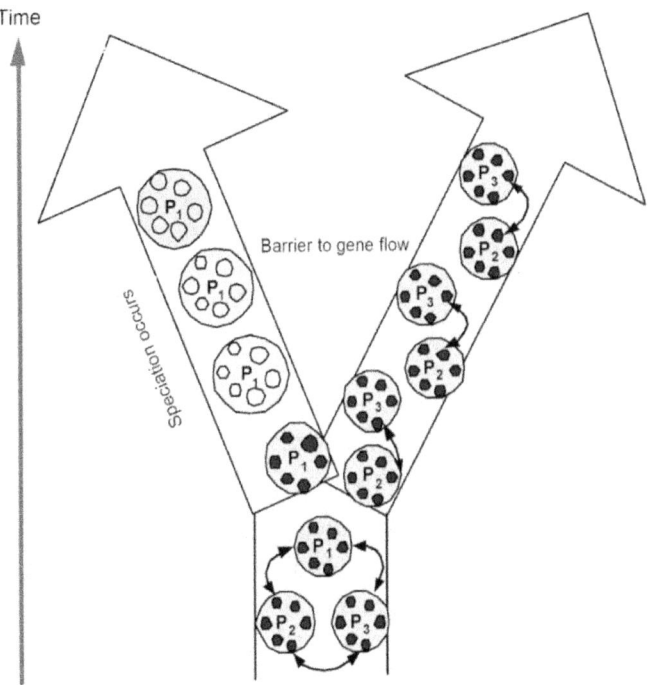

Figure 3.2 Origin of a new species due to reproductive isolation.

Reproductive isolation is any of the barriers that prevent or impede the organisms from two populations of the same species or from two different species from interbreeding and producing viable and fertile offspring. Reproductive isolation within the same species causes speciation.

51

> **Note.** Speciation explains the unity behind diversity of life. Different species give rise to diversity, whereas the characteristics that they share due to their common ancestry account for the underlying unity.

From the perspective of reproductive isolation, speciation is of two types.

3.2 Two Types of Speciation: Allopatric and Sympatric

Reproductive isolation (no gene flow) is necessary for speciation to occur. One way for the reproductive isolation to occur is the geographic separation of a population into two subpopulations, this is called allopatric speciation. However, reproductive isolation can also happen without geographic separation, this is called sympatric speciation. These two types of speciation are discussed in the following.

Allopatric Speciation: Allopatric speciation is the formation of new species due to the geographic separation between two populations of the same species. This happens if the geographic separation stops the gene flow among the population. For example, the different environments and different mechanisms of evolution can put two geographically separated populations of the same species on two different paths of evolution. Thus the gene pools of the two populations may change in different ways.

Sympatric Speciation: Sympatric speciation is the process in which a new species forms within the same population without the geographic split. This speciation occurs from barriers to the gene flow within the same population due to factors such as habitat differentiation, polyploidy, and sexual selection. *Habitat differentiation* refers to the situation when a population changes its habitat, and therefore its environment including resources. *Polyploidy* is the process in which an organism's genome acquires more than two sets of chromosomes. For example, polyploidy can result from an accident during cell division in which chromosomes fail to separate during gamete formation. Polyploidy occurs in animals rarely, but it is common in plants. *Autopolyploidy* is the polyploidy that occurs between organisms of the same species, and *allopolyploidy* is the polyploidy that happens between organisms of different species (Figure 3.3).

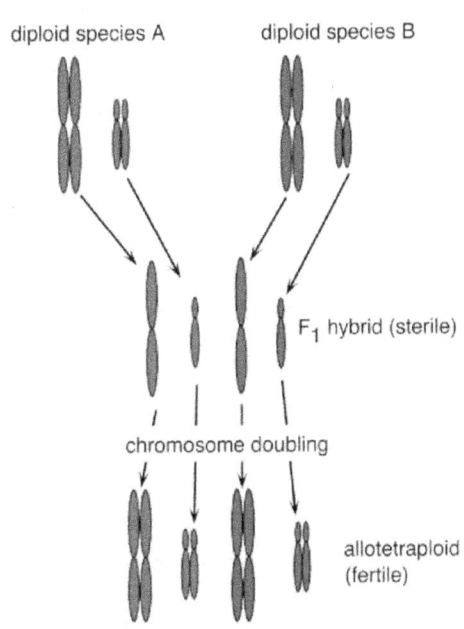

Figure 3.3 Hybridization event between the chromosomes from two different diploid species giving rise to allotetraploidy.

All of these different situations for speciation have one thing in common, the reproductive isolation, which is the necessary condition for speciation and is created by reproductive barriers, discussed next.

3.3 Reproductive Barriers

Reproductive barriers are required to create the reproductive isolation that is necessary for speciation and for maintaining separate identities of the existing species. A reproductive barrier is a barrier to gene flow, and it can exist between two different species so they can maintain their separate identities or within the same species that may lead to the emergence of a new species. All reproductive barriers can be grouped into two broad categories: prezygotic barriers and postzygotic barriers.

Prezygotic Barriers: *Prezygotic barriers* are the reproductive isolation mechanisms that prevent meeting, mating or fertilization. There are five different types of prezygotic barriers:

- **Spatial isolation:** This mechanism, also called *habitat isolation* or *ecological isolation,* is based on spatial separation of two species (or two populations of the same species) due to their different habitats. The different habitats may be in the same area or far apart, but the organisms from the two species do not meet due to spatial separation.

- **Temporal isolation:** This mechanism is based on separation by time and applies to two species that mate to reproduce at different times. For example, they may be mating different times of the day, different times (or seasons) of the year, or different years.

- **Behavioral isolation:** This mechanism is based on the differences among different species in pre-mating cues or rituals to attract mates. All the three mechanisms discussed so far prevent the attempt for mating.

- **Mechanical isolation:** This mechanism is based on morphological differences (physical incompatibilities) among different species that prevent the successful completion of mating when an attempt is made. All the mechanisms discussed so far prevent meeting, attempt to mate, or successful completion of mating.

- **Gametic isolation:** This mechanism is based on gamete incompatibility between two species that prevent fertilization after mating is successfully completed. For example, sperms of one species fail to fertilize the eggs of the other species.

If fertilization does happen between the gametes from the two species, *postzygotic barriers* come into play to prevent the development of viable and fertile offspring called hybrids.

Postzygotic Barriers: Postzygotic barriers are the reproductive isolation mechanisms that prevent the development of viable and fertile hybrid after a hybrid zygote has formed. They are discussed in the following:

- **Reduced hybrid viability:** This refers to physical problems that occur during the development of a hybrid species that may result in early death.

- **Reduced hybrid fertility:** This refers to the inability of hybrids to reproduce even if they survive.

- **Hybrid breakdown:** This refers to the non-viability and sterility of the hybrids after the first generation. In this case, first generation of hybrids may be viable and fertile, but when they mate with one another or with their parent species, they do not produce viable and fertile offspring.

> Caution! While gene flow within a species is required to hold the species together, reproductive isolation is necessary for new species to acquire and maintain their unique identities as species.

When reproductive isolation between existing species is compromised, hybrids are produced.

3.4 Role of Hybrids in Speciation

Hybrids are offspring produced as a result of mating between organisms from two different species. The offspring from two different true breeding types (for example, carrying two different alleles of the same gene) of the same species are also called hybrids, but here we will discuss hybrids from the mating between two different species, also called interspecific hybrids. For example, a mule (Figure 3.4) is a hybrid that results from mating between a male donkey and a female horse; and hinny is a hybrid that results from the mating between a male horse and female donkey.

Hybridization can take place between two mature species or subspecies, or between two populations that are on their way to become different species. To cover more situations in our discussion, we consider the latter.

Figure 3.4 Mules are interspecific hybrids and are sterile.

In this case, hybridization, depending on the reproductive barriers and the reproductive health of the hybrids, can lead to any of the following four possible results:

1. **New species from the hybrids with no gene flow from parents:** Hybrids are reproductively isolated from their parent species, the gene flow between the parent species and the hybrids stop, and the hybrids evolve into a new species.

2. **New species from the hybrids with gene flow from parents:** If the gene flow between hybrids and the parent species continue, one of the following three possibilities may result:

   - **Reinforcement of reproductive barriers:** If the reproductive isolation between two species is continually enforced and strengthened, production of hybrids will stop. IN this case, if the parent populations were on their way to becoming a new species, the speciation event will be completed. This can happen when the hybrids are less fit than their parent species.

   - **Fusion of species:** If the reproductive isolation between the parent species or populations is weakened, the two species may fuse into one. This can happen between two populations that were on their way to become distinctive species. In other words, the speciation process is reversed.

   - **Hybrid stability:** The parent species (or populations) continue their separate evolutionary paths, but also keep on forming hybrids. This happens when hybrids are able to survive and reproduce at rate compatible with that of the parent species.

> **Caution!** The time lapse between two speciation events on an evolutionary lineage depends on the environment and on other circumstances. It may be very short to very long and anywhere in between. Also depending on the situation and the functionality of genes, speciation may result from any number of genes from only a few to multiple.

In this chapter, we have worked with the definition of biological species based on the criterion of the capability of reproducing viable and fertile offspring. There are many other ways to define species based on other criteria. In the next section, we cover two other common definitions of species along with some other concepts.

3.5 More Definitions of Terms and Concepts

In this section, we define those important terms and concepts related to the topic at hand that are not already discussed in this chapter.

Biological Species: Biological species, as defined in Section 3.1, is the definition of species based on reproductive isolation that we have used so far in this chapter. We mention it here in context of other two definitions that we will explore. An advantage and credibility of this definition is that it is rooted in the process of speciation. A disadvantage is that it can only be applied to organisms that reproduce sexually and not asexually. Furthermore, it is not applicable to the fossil record. There is no way to determine the reproductive isolation in the fossil record in order to apply this definition there.

There are two other commonly used definitions of species: morphological species and phylogenetic species.

Morphological Species: A morphological species is defined by the unique set of its measurable anatomical features such as body shape. This is not absolutely different from the biological species concept because after all, most of the anatomical features (phenotypes) are based on genes. An advantage of this definition is that it can be applied to both sexual and asexual organisms. As a matter of fact, this is the definition that biologists have most commonly used so far to identify species. A disadvantage is the subjectivity involved in the anatomical criterion of this definition that sometimes gives rise to disagreements among scientists on which organism belongs to which species.

Phylogenetic Species: Phylogenetic species is the smallest group of organisms that share a common ancestor and is represented by a branch on the evolutionary tree of life. The advantage of this definition is that it is based on a widely applicable and precisely testable criterion. The only disadvantage is that it is a relatively new area. As it is a work in progress all the phylogenetic data is not yet available for many species.

Punctuated Equilibrium: This refers to the pattern found in the fossil record in which new species change the most at the time of their emergence from a parent species followed by a long static period, during which the species undergoes little or no change. The idea here is that once the biological divergence of a population from its parent species occurs, the speciation may occur rapidly, but it may take millions of years for this divergence to occur.

Biology of Evolution and Systematics

Adaptive Radiation: This is the process in which a lineage in the evolutionary tree rapidly gives rise to many new species through adaptations, which enable the new species to fill different niches offered by the environment. This may happen when a group of organisms lands in a geographic area where there is no competition or when a population evolves a novel trait that opens up a host of opportunities. The multitude of finch species on the Galápagos Islands is an example of adaptive radiation.

Dispersal and Vicariance: Dispersal and variance are two ways to create geographic isolation that facilitates allopatric speciation. Dispersal of a population (and hence alleles) happens when a fraction of the population moves to a new geographic location separate from the original location. Vicariance is the process in which a geological range in which a population lives splits into two disjointed parts creating a barrier for gene flow between two sections of the population on the two parts. Events of the continental split in the distant past, the emergence of a canyon, and the change of course of a river are some examples of vicariance.

Coevolution: Coevolution is the process in which two species evolve together by exerting evolutionary pressure on each other. There are some ecological interactions among species that cause them to coevolve. As a result of coevolution some species may become so interdependent that one cannot live without the other. Some examples of species coevolving are parasites and hosts, and predators and preys.

3.6 In a Nutshell

According to the biological definition of species, a species is one or more groups of organisms that have the potential to interbreed and produce viable and fertile offspring, and do not have the same potential to breed with other organisms of other groups. Gene flow within a species and reproductive isolation of the species from other species is necessary to keep the gene pool and hence the species together as one species. Reproductive isolation of a population is necessary in order to evolve into a new species different from the parent species. Reproductive isolation may occur with or without geographical separation of the population from the parent species. Speciation with geographic separation is called allopatric speciation, and speciation without geographic isolation is called sympatric speciation, which may occur within the parent species, for example, due to polyploidy events. Speciation may occur rapidly following a long period (millions of years) of no speciation event. This process is called punctuated equilibrium.

Because the biological definition of species is based on sexual reproduction, it cannot be applied to species that reproduce asexually. However, there are other definitions of species such as morphological species based on the anatomy of organisms and phylogenetic species based on evolutionary lineages. Morphological definition has been most widely used by scientists so far in order to identify most of the species, even though it is a somewhat subjective process.

The evolution of all the species on Earth has determined the biological history of life on Earth, a topic explored in the next chapter.

3.7 Review Questions

1. This is the process that gives rise to a burst of new species in a relatively short time to occupy a multitude of available niches in the environment without much competition. What is this process called?

   A. Coevolution

   B. Sympatric speciation

   C. Adaptive radiation

   D. Macroevolution

2. Which of the following is not the condition that will necessarily give rise to adaptive radiation?

   A. A strong storm causes some birds to stray from the mainland into an island that has many different resources but very little wildlife.

   B. Mass extinction of wildlife in a geographic area occurs.

   C. A plant population evolves a novel feature that opens up a host of opportunities to use in the environment.

   D. A small fraction of a population moves to a new area.

3. What is the process that facilitates speciation within the same population?

   A. Sympatry

   B. Allopatry

   C. Vicariance

   D. Dispersal

4. What is the fundamental condition for speciation?

    A. Allopatry

    B. Sympatry

    C. Polyploidy

    D. Reproductive isolation

    E. Gene flow

    F. All of the above

5. A plant with diploid chromosome number of 4 is crossed with another plant of diploid chromosome number of 6. The resulting polyploidy has how many chromosomes?

    A. Haploid number of 10

    B. Diploid number of 5

    C. Haploid number of 5

    D. Diploid number of 10

    E. Haploid number of 1

6. A scientist has discovered two species that are so interdependent they cannot live without each other. These species must have gone through a kind of evolution called:

    A. Convergent evolution

    B. Parallel evolution

    C. Joint evolution

    D. Coevolution

Biology of Evolution and Systematics

7. Plant species P_1 has a diploid number of 10. Plant species P_2 has a diploid number of 14. A new hybrid species, P_3, arises as an allopolyploid from A and B. The diploid number for species P_3 would probably be:

A. 14

B. 10

C. 12

D. 24

E. 48

8. Plant species P_1 has a diploid number of 10. Plant species P_2 has a diploid number of 14. A new hybrid species, P_3, arises as an allopolyploid from A and B. The gamete chromosome number for species P_3 would probably be:

A. 14

B. 10

C. 12

D. 24

E. 48

9. Which speciation concept is linked to the way speciation occurs?

A. Biological

B. Morphological

C. Phylogenetic

D. Ecologica

10. All of the following are postzygotic barriers to reproduction except:

A. Gametic isolation

B. Reduced hybrid viability

C. Reduced hybrid fertility

D. Reduced hybrid breakdown

3.8 Answer Key

1.	C	6.	D
2.	D	7.	D
3.	A	8.	C
4.	D	9.	A
5.	C	10.	A

Notes:

Q1. Sympatric speciation and macroevolution would be very broad answers. The correct answer to this specific situation is adaptive radiation.

Q2. Mass extinction creates the condition for adaptive radiation by reducing competition.

Q5. One parent contributes 2 chromosomes and the other parent contributes 3 chromosomes through gametes. These chromosomes cannot come together as homologous pairs. So, a haploid plant results with 2+3= 5 chromosomes.

Q7. Diploid number is the number of chromosomes in a somatic cell, 2n. This means the number of haploid chromosomes, n, in two gametes in this example will be 10/2=5, and 14/2=7. So, total number of haploid chromosomes = n = 7+5=12. Diploid numbers are always twice the haploid numbers. Therefore total number of diploid chromosomes = 24.

Q8. Gamete chromosome number is haploid chromosome number = n = 10/2 + 14/2 = 12

Chapter 4

History of Life on Earth

4.1 History of Life: Big Picture

The two expressions *history of life on earth* and *evolution of life on earth* are synonymous because life evolved. In other words, evolution defined the history of life on earth.

In terms of evolution, a large part of the history of life is written on sedimentary rocks in the form of fossils. A very brief summary of the information is collected from the fossil record by using techniques such as radiometric dating, as presented in Figure 4.1 and Table 4.1. The history of life is divided into three periods called eons: the Archaean, the Proterozoic, and the Phanerozoic. The Phanerozoic eon, as it is more crowded with evolutionary events as compared to other eons, is further divided into three sub-periods called eras: the Paleozoic, the Mesozoic, and the Cenozoic. The boundaries of these periods are marked by significant events: originations and extinctions. The important events in the history of life are listed below:

The Archaean Eon (4.6 bya-2.5 bya): The Archaean eon begins with the origin of Earth and ends with an increase in the levels of oxygen in the atmosphere created by cyanobacteria, a prokaryote. Life originated and prokaryotes evolved during this period.

Biology of Evolution and Systematics, by Paul Sanghera
Copyright © 2015 Infonential.

Biology of Evolution and Systematics

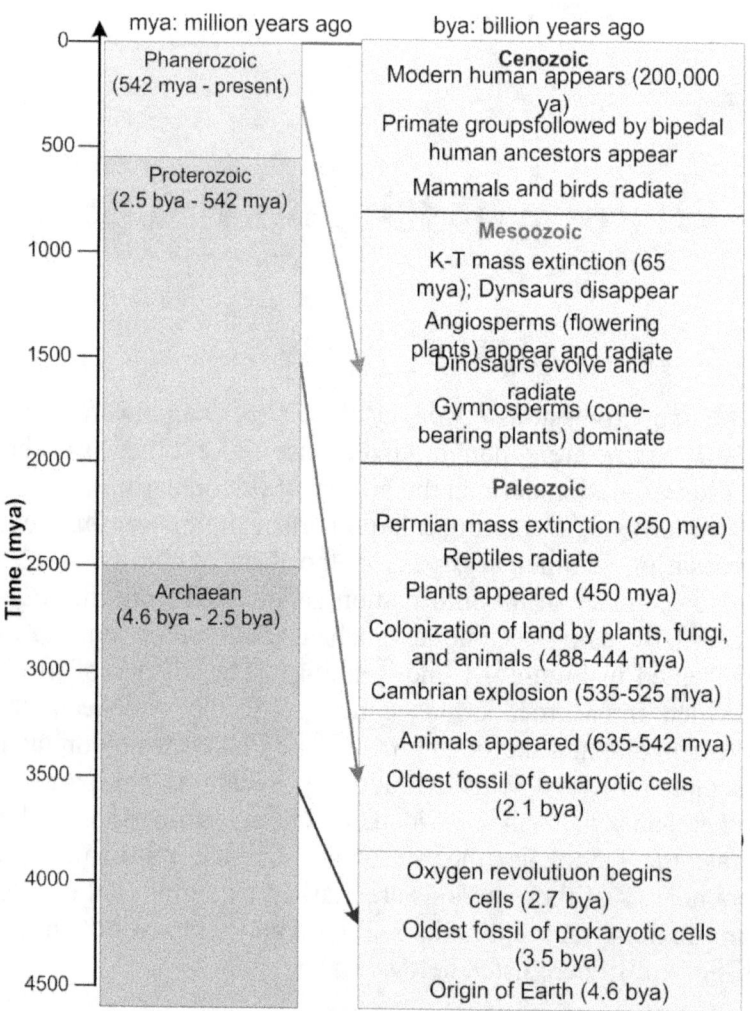

Figure 4.1 A brief history of life on Earth

The Proterozoic Eon (2.5 bya-542 mya): The boundary of the Archaean era and the Proterozoic era (2.7 bya to 2.3 bya) witnessed an oxygen revolution, the gradual rise in the amount of atmospheric oxygen. This was followed by the origin (evolution) of unicellular eukaryotes, multicellular eukaryotes, soft-bodied animals, and algae.

The Archaean eon and Proterozoic eon are referred to as pre-Cambrian periods, because they were followed by the Cambrian explosion which is discussed next.

The Phanerozoic Eon (542 mya – present): This eon began with the Cambrian explosion, an event that refers to a sudden increase and diversification in the number of animal groups such as species and phyla. The Cambrian explosion happened in the ocean. Following the Cambrian explosion was the move to land: first by fungi and plants, and then by animals. Early vascular plants, fish, and reptiles diversified. This period of the Phanerozoic eon is called the Paleozoic era (542 mya – 251 mya). It ended with a mass extinction called the *Permian* mass extinction, the biggest of all the mass extinctions. The Permian wiped out many species. For example, 96 percent of marine animal species and 70 percent of terrestrial vertebrate species became extinct.

This mass extinction opened the door for the second period of this eon, called the Mesozoic era (251mya – 65 mya), in which seed plants, especially the gymnosperms, diversified. In addition to being the age of the gymnosperms, the Mesozoic era also witnessed the age of dinosaurs (reptiles), which diversified as well. This era ended with a mass extinction called the Cretaceous (or K/T) mass extinction, which wiped out many groups of organisms including the dinosaurs, about half the marine phyla, and many plant species.

With the K/T mass extinction, came the current era called the Cenozoic era (65 mya – present), which has witnessed the age of the angiosperms (flowering plants), the age of birds, and the age of mammals.

Biology of Evolution and Systematics

Table 4.1 History of life on Earth, depicted by the fossil record.

Eon/Era	Time Range	Chronology of Events
Eon: Archaean	4.6bya – 2.5 bya	Origin of: Earth Life (protocells) Prokaryotes Oxygen revolution begins
Eon: Proterozoic	2.5 bya – 542 mya	Oxygen revolution continues Origin of: Eukaryotes Multicellular organisms Algae Animals
Eon: Phanerozoic	542 mya – present	
Era: Paleozoic Age of fish	542 mya – 251 mya	Non-vascular Plants Cambrian explosion: radiation of animals Move to land by plants Move to land by animals Origin of seedless vascular plants Origin of seed vascular plants Radiation of reptiles Permian extinction
Era: Mesozoic Age of: Gymnosperms (flowerless seed plants) Dinosaurs (reptiles)	251 mya – 65 mya	Seed plants (gymnosperms) dominate Dinosaurs appear and dominate Origin of angiosperms (flowering plants) K-T extinction: dinosaurs disappear
Era: Cenozoic Age of: birds, mammals, and angiosperms (flowering plants)	65 mya – present	Radiation of mammals Angiosperms dominate Origin of primate groups Origin of bipedal human ancestors, Ice Age, *Homo sapiens*, human

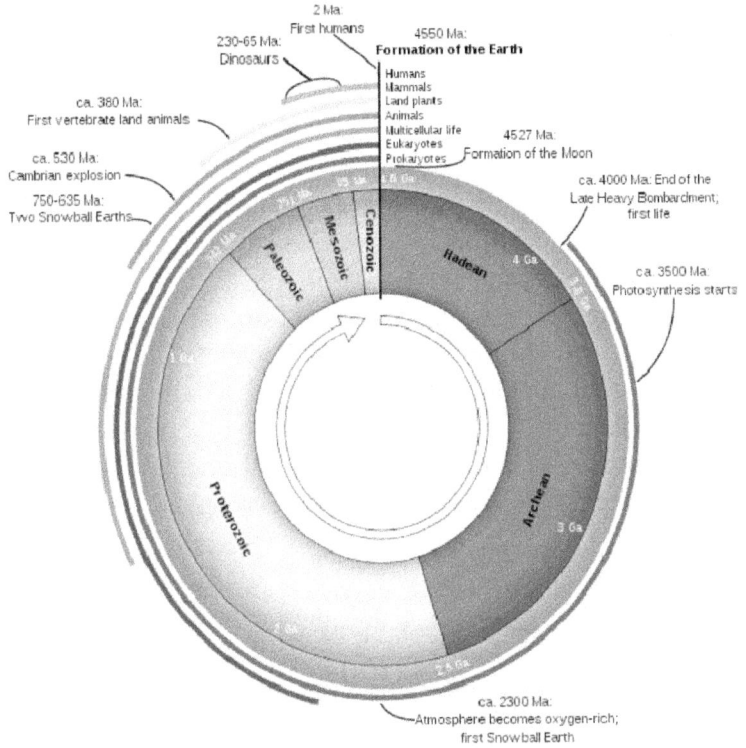

Figure 4.2 . Twelve-hour clock representation of the history of life on Earth.

Careful readers must have realized by now that most of the events that occurred during the evolution of life are crowded into the last half billion years, that is, the Phanerozoic eon. We, the human, arrived at the scene only about 200,000 years ago. If you scaled time from the origin of Earth until now into a 12 hour clock, as depicted in Figure 4.3, then the span of human existence would account for only a mere fraction of a second.

71

We, the late comers to the party of life on Earth, are indeed very curious. Among many other things, we want to know how life originated.

4.2 Origin of Life

Exactly when and how did life originally appear on Earth? This issue is not well settled yet. However, based on fossil record and experimental studies in physics, zoology, and biology; there is a general agreement among scientists that life began with simple cells, which came into existence through a spontaneous process, as depicted in Figure 4.3. This spontaneous process, governed by physical and chemical principles and the law of natural selection operating at micro-level, produced simple cells through the following four steps or stages:

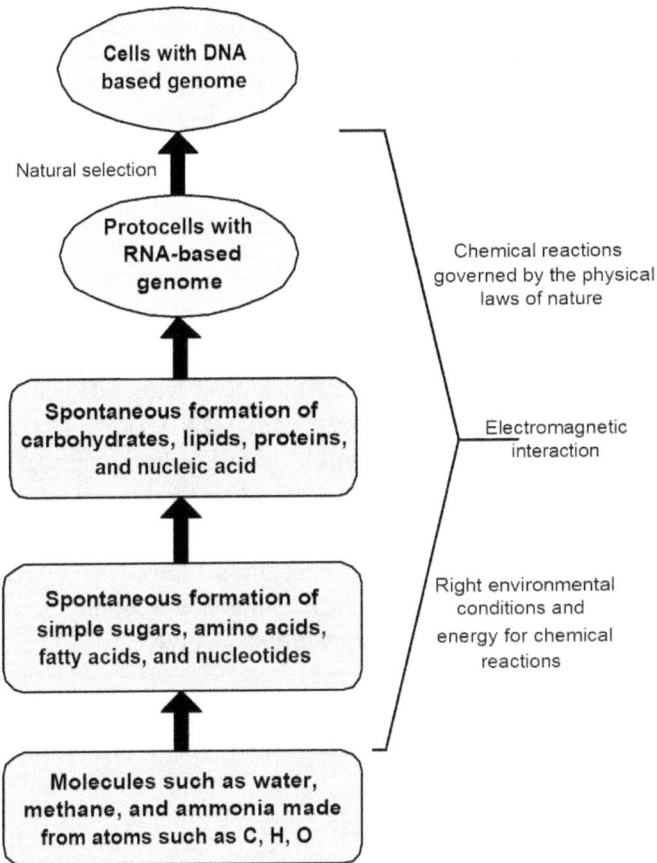

Figure 4.3 Spontaneous processes from which the life on Earth originated in form of cells, the fundamental building blocks of life.

1. **Micromolecules of life:** Abiotic physical and chemical processes synthesized fundamental organic molecules of life such as amino acids, nitrogenous bases (A, C, G, and T), and fatty acids.

2. **Macromolecules of life:** The micromolecules joined together to synthesize macromolecules of life: carbohydrates, lipids, proteins, and nucleic acids (DNA and

73

RNA). For example, multiple amino acid molecules joined to make a protein molecule.

3. **Protocells, the packages of life:** Physical and chemical processes put the molecules of life, the macromolecules, into a package that developed its own internal physical and chemical environment separate from the external environment.

4. **Self-replicating molecules of life:** The self-replication and metabolism capabilities, necessary for life, evolved inside the protocell. By using these capabilities, some molecules such as a nucleic acid (DNA or RNA) could replicate themselves, which would be necessary to replicate (divide) the protocell, and later, the cell.

Caution! Metabolism and replication are two essential conditions for life to arise, and the isolated environment created by the protocell helped these capabilities evolve.

After life originated, how did it get here and where it is today?

From Bacteria to Human: A Long Winding Road

In section 4.1 we overviewed the history of life. Starting with a protocell, discussed in section 4.2, life has evolved into incredibly diverse groups of organisms. Yet, looking closely with scientific scrutiny, there is an underlying unity behind this diversity due to a common origin. This is why scientists are able to classify all life into three clusters of closely related organisms, called domains: Archaea, Bacteria, and

Eukarya. Archaea and bacteria are collectively called prokaryotes. A prokaryote is unicellular organism with a relatively simple cell that contains a circular strand of DNA and is surrounded by a membrane and a wall. Unlike a eukaryotic cell, there are not any membrane bound organelles in a prokaryotic cell.

All the life we see around us with our naked eyes belongs to the domain Eukarya. A eukaryote could be a unicellular or a multicellular organism. A eukaryotic cell is more complex than a prokaryotic cell, as it contains membrane bound organelles including a nucleus that contains DNA. Evidence supports a theory called the *endosymbiont theory* according to which eukaryotes evolved from prokaryotes through a process called endosymbiosis, which is discussed further on in this chapter.

Today most organisms are aerobic, that is, they require oxygen to live. But the original prokaryotes were mostly anaerobic, that is, they made their cellular energy which is called ATP, without oxygen, because the oxygen level of the atmosphere at that time was very low. The first major evolutionary split of these prokaryotic cells gave rise to bacteria and the ancestor of archaeans and eukaryotes, which subsequently diverged into Archaea and Bacteria. The process of photosynthesis evolved in some groups of bacteria including cyanobacteria, which brought about the oxygen revolution around 2.5 bya. The rising levels of atmospheric oxygen due to the oxygen revolution helped evolve aerobic respiration, which is the process used among most organisms today to convert food into cellular energy, ATP.

We, the human are also eukaryotes. In addition to the discussion in this section, Figure 4.1 and Table 4.1 present a brief description of this history of life from prokaryotes to human.

75

4.3 General Pattern of the History of Life

The pattern of the history of life that emerges from the fossil record has three salient features:

1. Temporal Change. There have been different kinds of organisms living on earth at different times. Organisms that are living on Earth today are different from the organisms that lived in the past.

2. Extinction. Many kinds of organisms that lived in the past are now extinct.

3. Micro to macro. The change in the types of organisms over time has been generally the change from small to large, from simple to complex, from unicellular organisms to multicellular organisms.

These three observations are consistent with the theory of evolution.

4.4 More Definitions of Terms and Concepts

In this section, we define and discuss those important terms and concepts related to the topic at hand that are not already defined or adequately discussed.

The Endosymbiont Theory: The Endosymbiont Theory is based on endosymbiosis, which is an intimate and permanent symbiotic relationship wherein one organism lives inside another organism. According to the Endosymbiont Theory, which is well supported by evidence, the internal organism becomes an organelle of the cell in which it lived. This theory explains how, for example, a eukaryote evolved from a prokaryote when a larger prokaryote engulfed a smaller prokaryotic cell. The engulfed cell began receiving nutrients from the host cell and started dividing along with the larger cell. Gradually, it evolved into an organelle inside the host cell, and violá; we had a eukaryote. Another example is that of the mitochondria. The mitochondria evolved from a small aerobic prokaryote that entered a larger cell as an undigested food or as an internal parasite. Similarly, the ancestors of plastids, the photosynthesis related organelles, were originally photo-synthetic prokaryotes that entered the larger cells.

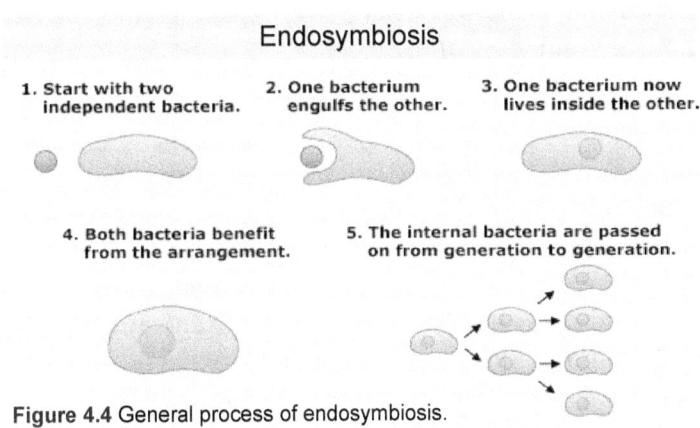

Endosymbiosis

1. Start with two independent bacteria. 2. One bacterium engulfs the other. 3. One bacterium now lives inside the other.

4. Both bacteria benefit from the arrangement. 5. The internal bacteria are passed on from generation to generation.

Figure 4.4 General process of endosymbiosis.

The general process of endosymbiosis is depicted in Figure 4.4.

The Plate Tectonics Theory: The plate tectonics theory is based on the verified fact that Earth's outer thin layer is composed of a number of distinct rocky plates floating on the semi-molten internal mantle. Our continents correspond to these moving plates. The plates (and hence the continents) move due to the processes occurring underneath within the mantle, and their motion is driven by well understood physical laws. Around 250 mya, this motion, called the continental drift, brought all of the continents together into one super continent called *Pangaea*. Then the continents subsequently separated and gradually moved to their current locations.

The continental drift explains many observations and patterns throughout the evolutionary history of life on Earth. For example, dinosaurs lived during the time of Pangaea. Therefore, we should be able to find their fossils all across the world; which we have. The continental drift has influenced the evolution of life by impacting the physical environment including climate.

Mass extinctions: As mentioned earlier in this chapter, mass extinctions have played an important role in the evolution of life on Earth. A mass extinction is defined as an event or a set of events that cause a sharp decrease in the diversity of life in terms of groups of organisms such as species. About 99 percent of documented species are now extinct, largely due to mass extinction events. We have already discussed two major mass extinctions, the Permian and K/T extinctions, in this chapter. There were three possible correlated reasons for both major mass extinctions: asteroid impact, increased tectonic activity such as volcanoes, and climate change such as changing ocean circulation patterns.

Effects of the mass extinctions on evolution: By eliminating certain species, a mass extinction drastically changes the path of evolution. At the expense of the extinct species, new species can evolve, in some habitats, as adaptive radiations, when there is less competition for resources.

4.5 In a Nutshell

Somewhere around four billion years ago primitive cells called protocells evolved from organic molecules such as nucleic acids and lipids. These protocells evolved into prokaryotic cells, unicellular organisms, by around 3.5 bya. Endosymbiosis helped the evolution of eukaryotes from prokaryotes around between 2.7 and 2.1 bya. By 900 mya, all major lineages including animals, fungi, and algae had evolved; plants evolved from algae. In addition to endosymbiosis, the oxygen revolution and later a number of mass extinctions shaped the journey of life on this evolutionary path. Here is the chronology of some of the important events in the history of life in the order they occurred:

1. Origin of prokaryotes

2. Oxygen revolution begins

3. Origin of eukaryotes

4. First (soft-bodied invertebrate) animals

5. Cambrian explosion

6. Age of fish

7. Fungi and plants move onto and

8. Animals move on to land starting with arthropods followed by amphibians

9. Permian extinction

10. Age of reptiles and seed plants

11. K-T extinction

12. Age of mammals and flowering plants

13. Origin of primate groups

14. Ice ages and origin of genus Homo

15. Origin of today's human

Therefore, starting from a single cell or a single type of cell, evolution has resulted in the splendid diversity of life we have today. At first glance, the task of studying such diverse life may seem impossible. However, recognizing the thread of evolution running through this diversity makes this impossible task possible. Scientists often begin the study of a complex and diverse system by first classifying it. This study technique applied on the diversity of life is called systematics, the subject of next chapter.

4.6 Review Questions

1. Starting with the origin of earth, which of the following is the correct chronological order, beginning from what appeared first?

   A. Prokaryotes, eukaryotes, plants, photosynthesis, oxygen revolution, animals.

   B. Prokaryotes, eukaryotes, animals, plants, photosynthesis, oxygen revolution.

   C. Eukaryotes, prokaryotes, animals, plants, photosynthesis, oxygen revolution.

   D. Prokaryotes, photosynthesis, oxygen revolution, eukaryotes, animals, plants.

2. Most scientists agree that life on Earth evolved in the following order:

   A. Monomer organic molecules, polymers of organic molecules, metabolism and replication, protocell, prokaryote.

B. Monomer organic molecules, polymers of organic molecules, protocell, metabolism and replication, prokaryote.

C. Polymers of organic molecules, monomer organic molecules, protocell, metabolism and replication, prokaryote.

D. RNA, DNA, polymers of molecules, metabolism and replication, protocell, prokaryote.

3. Which geologic period witnessed the age of fish?

A. The Proterozoic eon

B. The Cenozoic era

C. The Paleozoic era

D. The Mesozoic era

4. The Mesozoic era was:

A. The age of flowerless seed plants

B. The age of dinosaurs

C. The age of fish

D. A and B

E. A, B, and C

5. Which of the following came first in the history of life?

A. Unicellular eukaryotes

B. Prokaryotes

C. Unicellular protists

D. A and B

6. All eukaryotes have mitochondria, and only a few eukaryotes, the photosynthetic eukaryotes, have plastids. This means the following:

   A. Mitochondria evolved before plastids.

   B. Some cells that already had mitochondria engulfed a photosynthetic prokaryote.

   C. Mitochondria and plastids evolved in parallel.

   D. A and B

   E. None of the above

7. Life originated with the spontaneous assembly of complex organic molecules called molecules of life. This spontaneous process would not be possible if there was abundance of THIS substance in the atmosphere:

   A. Hydrogen

   B. Oxygen

   C. Nitrogen

   D. Ammonia

   E. A, C, and D

8. Which of the following process gave rise to the oxygen revolution?

   A. Aerobic respiration

   B. Anaerobic respiration

   C. Photosynthesis in bacteria

   D. Photosynthesis in plants

9. Scientific evidence indicates that the organelle mitochondria in eukaryotic cells evolved from:

   A. Anaerobic bacteria

   B. Aerobic bacteria

   C. Another organelle

   D. Cyanobacteria

10. Scientific evidence indicates that the organelle chloroplast in some eukaryotic cells evolved from:

   A. Anaerobic bacteria

   B. Aerobic bacteria

   C. Another organelle

   D. Cyanobacteria

4.7 Answer Key

1.	D	6.	D
2.	B	7.	B
3.	C	8.	C
4.	D	9.	B
5.	B	10.	D

Notes:

Q7. Oxygen is very reactive and would have reacted with the organic molecules as soon as they were formed. This is one of the reasons why we do not witness any spontaneous appearance of life in this oxygen-rich environment today.

Q10. Chloroplast and other plastids, the photosynthetic organelles, most likely evolved from photosynthetic bacteria engulfed by the larger cells.

Chapter 5

Taxonomy and Phylogenetics

5.1 Systematics: The Big Picture

Systematics is a sub-field of biology focused on the study of biodiversity by classifying organisms and determining their evolutionary relationships. A general methodology used in all science disciplines is to group or classify things based on various criteria such as shared characteristics. Systematics also involves classifying organisms into groups. Historically, there are two techniques to classify organisms: taxonomy and phylogenetics. The big picture of system-atics

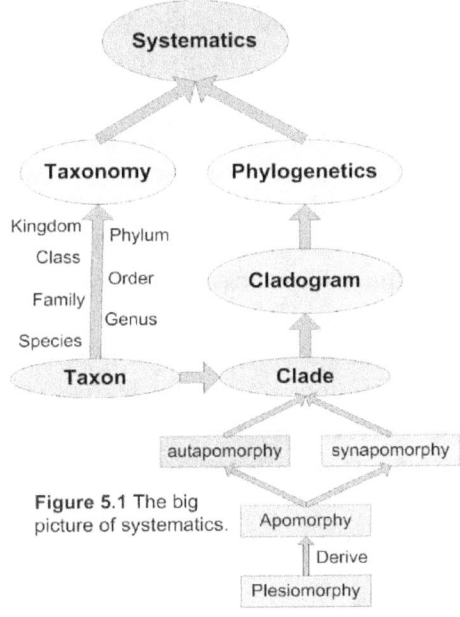

Figure 5.1 The big picture of systematics.

is presented in Figure 5.1. The terms and concepts referred to in this figure will become clear as you venture through this chapter.

Biology of Evolution and Systematics, by Paul Sanghera
Copyright © 2015 Infonential.

> Note. Due to his religious beliefs, Linnaeus, in spite of the fact that he originated taxonomy, could not see in it what Darwin was able to see: the classification of organisms reflected the evolutionary relationships among them.

Taxonomy, the naming and classification scheme, was developed even before the theory of evolution was proposed.

5.2 Taxonomy

Taxonomy is the system developed by a Swedish physician Carolus Linnaeus (1707-1778) to classify the diverse forms of life. As shown in Figure 5.2, all forms of life are sorted into taxonomic groups that are organized into hierarchical super-type subtype relationships, specifically species, genus, family, order, class and kingdom. The relationships between these groups are also called generalization-specialization or parent-child relationships. In this relationship, a subtype is called a kind of super-type. For example, a human is a kind of Sapiens (species), which is a kind of Homo (genus), which in turn is a kind of Hominidae (family), and so on. In figure 5.2, examples of each group, called taxon, at each level of classification is given for human. The groups of organisms at each level of classification are commonly called taxa (singular: *taxon*), and are explained below.

Species: This is the most specialized (the lowest subtype) in the taxonomy. We belong to the species sapiens, written as a binomial: *Homo sapiens*.

Genus: Two or more similar species are grouped into a super-group called genus (plural: *genera*). We belong to the genus homo.

Family: Two or more similar genera are grouped together into a super-group called family. We belong to the family hominidae.

Order: Two or more similar families are grouped together into a super-group called order. We belong to the order primates.

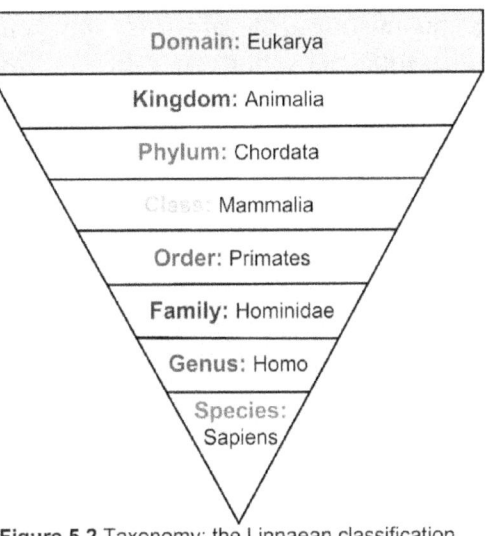

Class: Two or more similar orders are grouped together into a super-group called class. We belong to the class mammalia.

Figure 5.2 Taxonomy: the Linnaean classification of organisms; human used as an example.

Phylum: Two or more similar classes are grouped together into a super-group called phylum (plural: phyla). We belong to the phylum chordata.

Kingdom: Two or more similar phyla are grouped together into a super-group called kingdom. We belong to the kingdom animalia.

Domain: Kingdoms are organized into a super-group called domain, the highest level of classification. Classification into domains is a relatively recent development based on studies in cellular/molecular biology including genetics. Because cells are the fundamental building blocks of life, it only makes sense to classify organisms based on cells. Organisms classified this way, fall into three domains: Bacteria, that consists of the most diverse and widespread unicellular organisms made of prokaryotic cells; Archaea, that consists of organisms that are also unicellular and made of prokaryotic cells but are different from bacteria in many aspects such as the ability to live in extreme environments; and Eukarya, that consists of organisms made of eukaryotic cells and can be unicellular or multicellular.

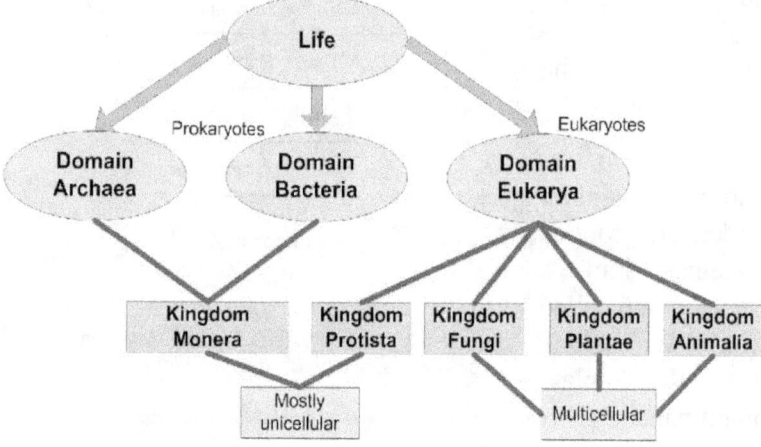

Figure 5.3 Relationship between domains and kingdoms.

As shown in Figure 5.3, the kingdom Monera consists of prokaryotes (which are mostly unicellular) and is split into two domains: Bacteria and Archaea. Domain Eukarya consists of four kingdoms: three multicellular kingdoms namely Fungi, Plantae, and Animalia; and one mostly unicellular kingdom named Protista. We, the human, belong to the domain Eukarya.

As mentioned earlier, taxonomy was originally developed before the discovery of evolution. The most fundamental unit in a taxonomic system is species. The evolutionary history of a species or a group of species is called phylogeny, explored in a discipline called phylogenetics.

> **Binomial.** A binomial is a scientific name in the format developed by Linnaeus according to which a species is named in two parts: the second part is the italicized name of species (specific epithet) in lower case, and the first part is the italicized name of the genus to which the species belong, with first letter in upper case. For example, the binomial for us, the human is *Homo sapiens* because we belong to the sapiens species and homo genus.

5.3 Phylogenetics

Phylogenetics is the study of phylogeny, the evolutionary history of organisms or groups of organisms. One very important approach in phylogenetics is called cladistics, introduced in 1950 by a German entomologist named Willi Hennig in which organisms are classified based on their common ancestry. This naturally gives rise to an evolutionary tree called cladogram.

Cladogram: A cladogram is an evolutionary tree built according to the rules of cladistics in which each node (branch point) represents a speciation event in which an ancestral species splits into two daughter species. A cladogram is a hypothesis of evolutionary relationships determined by phylogenetics and usually suggested by data.

Clade: A group of taxa in a cladogram that includes one ancestral taxon and all of its descendents. Each clade is defined by a unique set of features that the taxa in the clade possess and any other group of taxa lack. A clade is also called a monophyletic group as opposed to paraphyletic group and polyphyletic group. A *paraphyletic group* is a group of taxa that consists of one ancestor and some (not all) of its descendents, whereas a polyphyletic group is a group of taxa that does not have a common ancestor but does have at least one common or similar characteristic shared by all taxa in the group, which may have appeared through convergent or parallel evolution.

Derived feature: Consistent with evolution, cladistics assumes that certain traits (also referred to as features or characteristics) of organisms are heritable, that is, can be passed down to the next generation. Further, in line with evolution, cladistics assumes that heritable traits are subject to change over time. If the change in a trait occurs, such as from *laying shelled egg* to *giving live birth*, we refer to the original state, the ancestral state and the changed (new) state, the derived state or the derived characteristic. The ancestral state or feature is also called plesiomorphy, and the derived feature as apomorphy.

Taxon: Taxon is a group of organisms defined at any level of taxonomy or classification. For example, any group from species through kingdom can be considered as a taxon. In cladistics, as Hennig proposed, a taxon should be grouped together based on a unique set of features that only the particular group (taxon) posses and the taxa outside of the group lack.

Autapomorphy: When there is a derived trait that is unique to a single taxon, the trait is called autapomorphy. If over generations, the characteristic changes (for example a long beak into a short beak), the old state of the characteristic is called ancestral state or feature and the new (changed) state is called derived state or derived feature.

Synapomorphy: Any derived feature that is shared by two or more taxa and hence helps identify its clade is known as synapomorphy.

Apomorphy: The derived state or feature in a taxon. In other words, autopomorphies and synapomorphies are collectively called apomorphies.

Plesiomorphy: A characteristic of a taxon in its ancestral state is called plesiomorphy.

Caution! Regardless of what groups are represented by taxa in a cladogram (species, genus, or order etc.), a node in the cladogram always represents a speciation event. This is because the origin of any taxon can always be traced back to a species, as all taxa originated as species.

We have defined many terms in this previous section. They will become clearer if you work through the example presented in the next section.

5.4 Explanation Through Example

Problem 5.1 Consider the following cladogram:

Biology of Evolution and Systematics

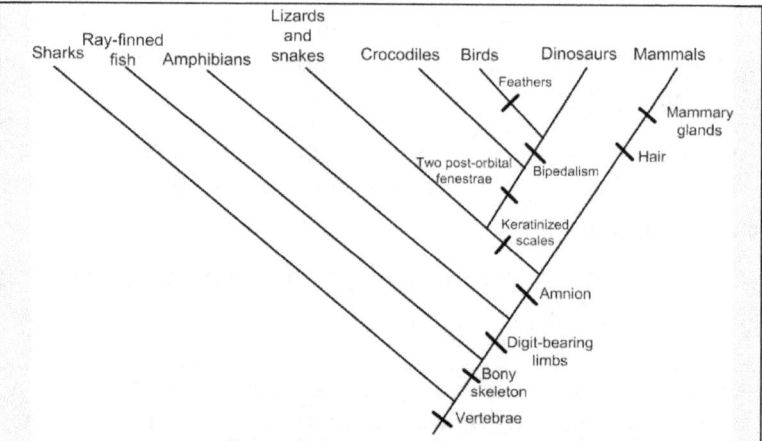

Figure 5.4 Taxonomy: A cladogram of vertebrates

Answer the following questions:

Q1. What synapomorphy puts birds, crocodiles, and dinosaurs into a clade of their own?

A1: Two post-orbital fenestrae.

Q2. What are the plesiomorphies of amphibians, lizards and snakes, crocodiles, birds, dinosaurs, and mammals?

A2: bony skeleton, and vertebrae.

Q3. What is the synapomorphy of amphibians, lizards and snakes, crocodiles, birds, dinosaurs, and mammals?

A3: Digit-bearing limbs.

Q4. List all the autopomorphies in the cladogram.

A4: Hair and mammary glands for mammals, and feathers for birds.

Q5. Are crocodiles more closely related to birds or dinosaurs?

A5: Crocodiles are equally closely related to birds and

dinosaurs because crocodiles share their most recent ancestor with both birds and dinosaurs.

Q6. What are the birds most closely related to?

A6: Dinosaurs.

Q7. What is the synapomorphy for all the taxa in the cladogram?

A7: Vertebrae

Caution! Note that apomorphy and plesiomorphy are relative terms; a plesiomorphy for one clade may be an apomorphy for another. For example, in Figure 5.4, the existence of two post-orbital fenestrae is an autapomorphy (synapomorphy in this case) for crocodiles, birds, and dinosaurs; whereas it is a plesiomorphy for birds and dinosaurs.

5.5 In a Nutshell

All the diverse organisms on Earth are united by their evolutionary history, the phylogeny. The study of phylogeny is called phylogenetics, and it is a part of systematics, the discipline of biology in which scientists name and classify organisms and study evolutionary relationships among them. One very important approach in phylogenetics is called cladistics, introduced in 1950 by a German entomologist named Willi Hennig in which organisms are classified based on their common ancestry. This naturally gives rise to an evolutionary tree called a cladogram.

93

Cladistics is based on two assumptions consistent with the theory of evolution:

1. **Branching evolution:** The history of life is characterized by an evolutionary branching pattern. A node represents a speciation event where an ancestral species (branch) produced two daughter species (branches). Such a branching pattern is also referred to as dichotomous splitting pattern.

2. **Derived features:** Evolutionary relationships are determined by features that are heritable and are subject to change over time, that is, new features are derived from old (ancestral) features.

5.6 Review Questions

1. A scientist is studying a group of coyotes that belong to the species latrans and the genus canis. How would the scientist write down the scientific (binomial) name for these animals?

   A. *Latrans canis*

   B. *Canis Latrans*

   C. Canis latrans

   D. *Canis latrans*

2. Modern systematics is based on the ideas of:

   A. Darwin, Linnaeus, and Hennig

   B. Darwin, Linnaeus, and Mendel

   C. Darwin and Mendel

D. Darwin, Weinberg, and Hardy.

3. Prokaryotes belong to _____ kingdom and _____domain(s):

   A. Monera, Archaea and Bacteria

   B. Monera, Bacteria

   C. Animalia, Archaea and Bacteria

   D. Protista, Bacteria

4. Which kingdom consists of mostly unicellular eukaryotes?

   A. Monera

   B. Prokarya

   C. Protista

   D. Animalia

5. Which of the following is the correct order in Linnaean classification?

   A. Kingdom, Class, Phylum, Order, Family, Genus, Species.

   B. Kingdom, Phylum, Class, Order, Family, Genus, Species.

   C. Kingdom, Phylum, Class, Family, Order, Genus, Species.

   D. Species, Genus, Family, Class, Order, Phylum, and Kingdom.

6. A clade is a:

   A. Polyphyletic group

   B. Paraphyletic group

C. Monophyletic group

D. B and C

E. None of the above

7. Which of the following is true in phylogeny?

A. Derived traits are always more complex than the ancestral traits.

B. A monophyletic is a group of taxa that consist of a common ancestor and all its descents.

C. An organism modifies its ancestral state to a derived state to adapt to the environment.

D. A polyphyletic group is also called a clade.

5.7 Answer Key

1. D	4. C
2. A	5. B
3. A	6. C
	7. B

Notes:

Q4. There is no kingdom named prokarya.

Q5. You can memorize this order of K, P, C, O, F, G, and S by easily remembering this phrase: *Kings Play Chess On Fine Glass Stools.* If you want to include domain (the highest level), the phrase becomes: *Dumb Kings Play Chess On Fine Glass Stools.*

Chapter 6

Evolution of the Monera Kingdom

6.1 The Monera Kingdom: Big Picture

The taxonomic rank kingdom is the second broadest category, second only to domain. Based on molecular data, it is now well established that all living organisms belong to three domains: Bacteria, Archaea, and Eukarya. While a few kingdoms have been organized into the domain Eukarya, the kingdom Monera has life forms that fall into two domains, the domain Bacteria and the domain Archaea. Bacteria and archaea are prokaryotes, which are mostly unicellular, whereas the domain Eukarya includes both unicellular and multicellular organisms. The kingdoms Animalia, Plantae, Fungi, and Protista are all parts of the domain Eukarya. Figure 6.1 depicts this picture of life. As Figure 6.1 indicates, archaea are closer to eukarya than to bacteria. This is also reflected in Table 6.1, where the features of bacteria, archaeans, and eukaryotes are compared.

Caution! Although eukaryotes can be both unicellular and multicellular, most eukaryotic lineages or species are unicellular.

Biology of Evolution and Systematics, by Paul Sanghera
Copyright © 2015 Infonential.

Biology of Evolution and Systematics

Table 6.1 A comparison of the three domains of life.

Feature	Bacteria	Archaea	Eukarya
Membrane-enclosed nucleus and organelles	Absent	Absent	Present
Circular chromosome	Present	Present	Absent
Histones (alkaline proteins) associated with DNA	Absent	Present in some species	Present
Introns in DNA	Very rare	Present in some genes	Present in many genes
RNA polymerase	One kind	Several kinds	Several kinds
Membrane lipids	Unbranched hydrocarbons	Some branched hydrocarbons	Unbranched hydrocarbons
Peptidoglycan in cell wall	Present	Absent	Absent
Growth at extreme temperature (> 100°C)	No	Some species	No

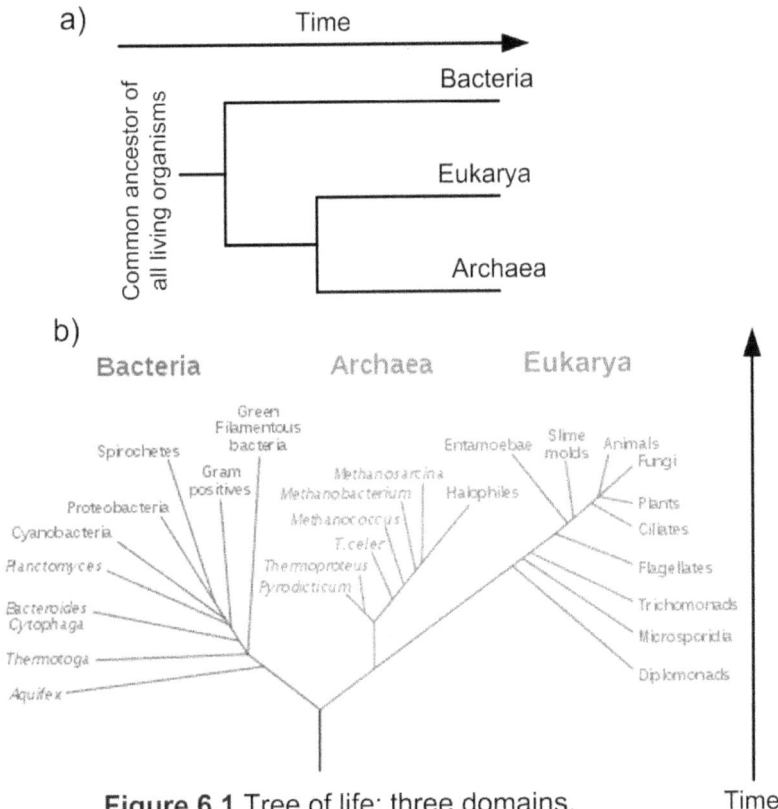

Figure 6.1 Tree of life: three domains.

Fascinating Fact! Prokaryotes outnumber any other type of organisms on Earth. For example, the number of bacteria in your mouth alone outnumbers all the people living on the planet Earth.

Because the Kingdom Monera consists of two out of the three domains of life, it is appropriate here to discuss the evolution of the three domains of life.

6.2 Evolution of Three Domains of Life

Figure 6.2 presents a more detailed version of Figure 6.1 part a). This tree that is supported by data indicates that the first major divergence or branching in the history of life occurred when the ancestors of Archaea and Eukarya separated from prokaryotes. Prokaryote lineage evolved into bacteria, whereas the organism in the lineage that split from prokaryotes evolved into Eukarya and Archaea. The gene transfer events from prokaryotes to this lin-eage contributed to this evolution as depicted in Figure 6.2. Such a transfer of

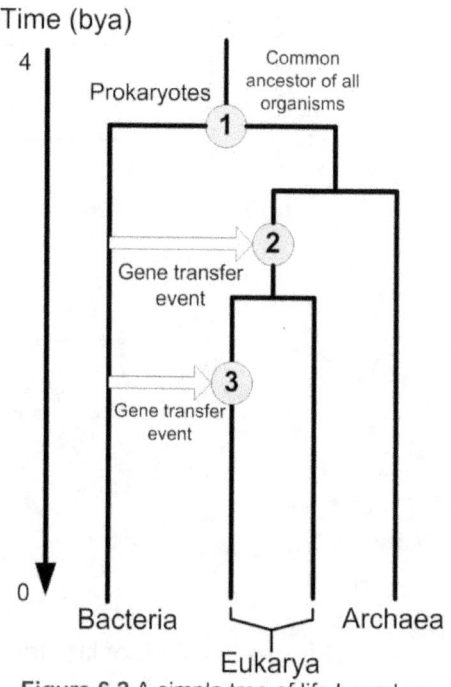

Figure 6.2 A simple tree of life based on three domains

genes between genomes of different lineages is called *horizontal gene transfer* or *lateral gene transfer*. For example prokaryotes, can directly inject their genes into another microbe of different lineage through a sexual act. The injected genes will become part of the host microbe, which will pass them down to its offspring, a process called *vertical gene transfer*. As illustrated in Figure 6.2, while gene transfer may have contributed to the split into domains, it also may have contributed to the split of a lineage into two lineages within the same domain.

Based on data collected for decades, scientists have developed a timeline estimate for the evolution of three domains of life:

1. **Prokaryotic cells:** The first living cell evolved around 3.8 billion years ago (bya). All cells and organisms evolved from that were prokaryotes to begin with.

2. **Bacteria:** Around 3.2 bya there was a divergence in the prokaryotic lineage giving rise to bacteria and to the common ancestors of archaeans and eukaryotic cells.

3. **Archaeans and eukaryotic cells:** The ancestors of archaeans and eukaryotic cells diverged into two separate lineages around 3.0 bya.

4. **Photosynthesis and aerobic respiration evolve:** Around 2.4 bya, cyanobacteria supporting photosynthesis appeared. Photosynthesis changed the Earth's atmosphere by continually adding oxygen to it.

5. **Emergence of cell nucleus:** The number of genes and the overall size of cells that would eventually becomes eukaryotic cells, kept increasing. Between 3.0 bya and 2.0 bya, the nuclear membrane emerged enclosing the genome and giving rise to the cell nucleus. Eukaryotic cells, the cells containing membrane-bound organelles such as a nucleus in them, appeared around 1.9 bya.

6. **Emergence of mitochondria:** By 1.2 bya, a would-be eukaryotic cell engulfed a small prokaryotic cell, a process called endosymbiosis, as discussed in Chapter 4. The smaller internal cell eventually became mitochondrion, an

organelle that carries cellular respiration in eukaryotic cells.

7. **Multicellular organisms:** Sexual reproduction and simple multicellular organisms appeared around 1.2 bya.

8. **Emergence of Chloroplasts:** Between 1.5 bya and 1.0 bya an evolving eukaryotic cell engulfed a cyanobaterium, a bacterium capable of performing photosynthesis, which evolved into chloroplast, an organelle that serves as the site for photosynthesis in plants and photosynthetic protists.

Because all prokaryotes belong to Kingdom Monera, they have similar anatomic and physiological features.

6.3 Kingdom Monera: Anatomic and Physiological Features

Some anatomic (Figure 6.3) and physiological features of prokaryotes (Kingdom Monera) are described in the following:

Flagella: As shown in Figure 6.3, a flagellum (plural: flagella) is a tail-like long appendage projecting from a cell, and is specialized for locomotion. Although flagella are mostly found in prokaryotic cells, some eukaryotes also have flagella. Anatomical and molecular studies suggest that flagella of bacteria, archaeans, and eukaryotes are analogous structures and not homologous.

Nucleoid: This is the region in a prokaryotic cell where DNA is concentrated. Unlike the nucleus of a eukaryotic cell, the nucleoid of a prokaryotic cell is not bounded by a membrane.

Plasmids: A small and usually circular-shaped DNA molecule in a prokaryotic cell that is separate from the DNA in the nucleoid is a plasmid. The plasmid contains accessory genes, for example, some bacteria have what is called R-plasmids, which carry the genes that code for enzymes that make bacteria resistant to antibiotics. Plasmids are also found in some eukaryotic cells such as yeast.

Ribosomes: A ribosome is a structure composed of rRNA and proteins, and is the site for protein synthesis in a cell. Both prokaryotes and eukaryotes have ribosomes in their cells.

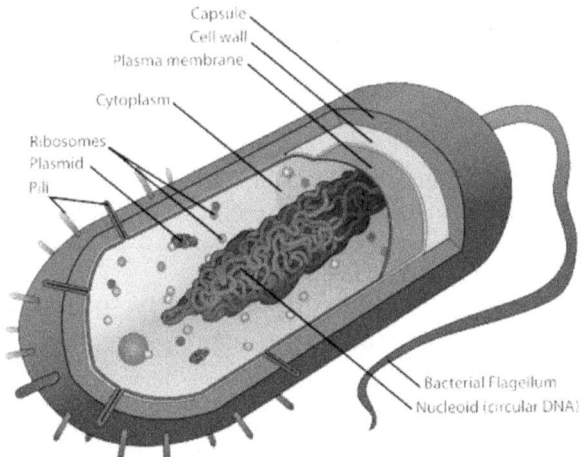

Figure 6.3 Illustration of a typical prokaryotic cell.

Binary fission: This is the process of asexual reproduction in which a cell (unicellular organism) splits into two, which is how prokaryotes reproduce. In prokaryotes, neither meiosis nor mitosis is needed in binary fission. However, in eukaryotes, binary fission does use mitosis. Many prokaryotes can typically divide every 1-3 hours. For example, *E. Coli* can divide every 20 minutes under favorable conditions.

Capsule: In many prokaryotic cells, a well defined sticky layer of polysaccharides or proteins surround the cell wall. This layer is called a capsule. The capsule protects the cell and helps it to adhere to other entities such as other cells and substrates.

Endospore: An endospore is a thickly coated, temporarily non-reproductive cell, which is resistant to harsh environments. This is essentially the dormant state which certain bacteria can reduce to. Endospores enable bacteria to lie dormant for an extended period, even centuries long, under harsh conditions and then revive when the conditions are right.

Fascinating Fact! Bacteria can survive harsh conditions for centuries by falling into the dormant state called an endospore and then reviving themselves when the conditions are right.

Rapid adaptation: Prokaryotes are highly successful due to three characteristics: small size, short generation time, and simple binary fission reproduction. These three properties lead to huge populations with the capability to respond quickly to all kinds of environment, resulting in rapid adaptations and evolution. Short generational time leads to high temporal rate of mutations, which leads to greater genetic diversity, which in turn leads to rapid evolution and adaptations, which results in greater overall reproduction fitness.

Note. Three reproduction features that contribute to the relatively large population of prokaryotes are: small size, binary fission, and short time between generations.

Metabolic adaptation: As discussed in Section 6.4, metabolic adaptations in prokaryotes are much larger in number and more diverse than in eukaryotes.

Sources of variation: In prokaryotic populations, mutations are the major source of variations. However, variations or diversity also comes from *genetic recombination*, which is the process of combining the DNA from two different sources. Genetic recombination in prokaryotes is of three types: transformation, transduction, and conjugation. In transformation, a prokaryote takes in foreign DNA from its surroundings, whereas transduction refers to the process in which a virus injects the foreign DNA into a prokaryote. Conjugation is the process in which one prokaryote transfers some of its DNA into another prokaryote directly through a structure called a sex pilus. Conjugation is a mechanism of horizontal gene transfer.

The F factor: The F stands for fertility. The F factor refers to the segment of a DNA in a prokaryote that enables it to form a sex pilus in order to carry on conjugation.

Biofilm: Biofilm is a surface coating formed by a colony of prokaryotes adhering to one another. Some types of biofilm are more commonly known as slime, even though not all slimes are biofilms. The slimy coating on a fallen log of wood and on your teeth are examples of biofilms. A biofilm is made by one or more prokaryotic species that get involved in metabolic cooperation, a process during which some channels in a biofilm help nutrients to reach cells in the interior and facilitate the expulsion of wastes from the biofilm.

> **Caution!** Do not underestimate prokaryotes. Even though they have structurally simpler cells than those of eukaryotes, they are not less evolved or primitive on the evolutionary scale. In light of their simplicity, they have incredible genetic diversity.

Capsules and endospores discussed in the list above are two important adaptations that help prokaryotes to survive in harsh environments.

The genetic diversity in prokaryotes has given rise to their diversity at the organism level such as nutritional and metabolic diversity. This diversity has helped them to outnumber and outmass any other form of life on this planet.

6.4 Diversity and Ubiquity of Prokaryotes

Prokaryotes, the organisms that make up the kingdom Monera, certainly made use of being first in the history of evolution of life on this planet. They have diverse metabolic and other adaptations, and as a result they have proliferated and are present almost everywhere on this planet. Their metabolic diversity in terms of nutritional modes is presented in Table 6.2.

Table 6.2 Metabolic diversity among prokaryotes in terms of major nutritional modes.

Nutritional Mode	Carbon Source	Energy Source	Organism Types
Autotroph: Chemoautotroph	CO_2 in some form: CO_2, HCO_3^-, etc.	Inorganic chemicals such as Fe^{2+}, H_2S, S, NH_3, etc.	Certain prokaryotes such as methanogens and halophiles
Photoautotroph	CO_2 in some form: CO2, HCO_3^-, etc.	Light such as sunlight	Plants, photosynthetic prokaryotes such as cyanobacteria, some protists such as algae.
Heterotroph: Chemoheterotroph	Organic nutrients such as glucose	Organic compounds such as glucose	Many prokaryotes, protists, animals, fungi, and some plants
Photoheterotroph	Organic nutrients such as glucose	Light such as sunlight	Certain prokaryotes such as heliobacteria and green non-sulfur bacteria

Photosynthesis and aerobic respiration first evolved in bacteria, a prokaryotic domain. As table 6.2 illustrates, there are prokaryotes corresponding to all kinds of nutritional modes: photoautotrophs, chemoautotrophs, photoheterotrophs, and chemoheterotrophs:

- **Photoautotroph:** A photoautotroph is an organism that uses some form of carbon dioxide as a carbon source and light such as sunlight as an energy source. Photoautotrophs carry out photosynthesis.

- **Chemoautotroph:** A chemoautotroph is an organism that uses some form of carbon dioxide as a carbon source and inorganic chemicals such as sulfur, hydrogen sulfide, ammonia, and ferrous iron as an energy source.

- **Photoheterotroph:** A photoheterotroph is an organism that uses organic compound such as glucose as a carbon source and light such as sunlight as an energy source. Photoheterotrophs carry out photosynthesis.

- **Chemoheterotroph:** A chemoautotroph is an organism that uses organic compounds such as glucose as a carbon source and also as an energy source.

The qualifiers *photo* and *chemo* refer to the energy source, whereas the qualifiers *auto* and *hetero* refers to the carbon source.

Although some bacteria do cause diseases, it is a common misconception that all bacteria cause diseases, and that is all they do. But bacteria are so much more. As a matter of fact, bacteria, or prokaryotes in general, play an important role in maintaining life in the biosphere.

6.4 Roles of Prokaryotes in Biosphere

All life on this planet depends directly or indirectly on sun, water, and nutrients such as carbon, hydrogen, nitrogen, and phosphorous. Water and nutrients are cycled through local and global cycles called biogeochemical cycles. Prokaryotes play key roles in these nutrient cycles on which life depends. They play their role in many ways including decomposition and mutualism explained in the following.

Decomposition: Decomposition is the process of breaking down dead organisms or body parts and the organic waste of living organisms in order to extract mostly inorganic nutrients such as nitrogen and phosphorous. These nutrients are cycled through the biogeochemical cycle to support life, that is, for the use of other organisms. Some prokaryotes perform this task and therefore are called decomposers or detritivores.

Mutualism: Mutualism is an interspecific interaction between two species that benefits both participating species. Many prokaryotic species form mutualistic relationships with other species. These relationships sustain the life of the participating organisms from both species. For example, the human intestine hosts hundreds of species of bacteria that are in a mutualistic relationship with us. They help our digestion system to process the food we eat, and in return they live on (digest) the food that the intestine cannot digest. We depend on these prokaryotes.

Fascinating Fact! The total number of prokaryotic cells in the intestine of a human is about ten times the total number of human cells in the body.

Biology of Evolution and Systematics

Medicine, Research, and Technology: Largely due to their diverse metabolic capabilities, prokaryotes are widely used in the fields of medicine, research and technology, as well as in the food industry. For example, some bacteria are used to make cheese and yogurt from milk. Other bacteria play an important role in the biotechnology industry, for example, *E. coli* is used to produce insulin, which treats certain types of diabetes. *E. Coli* is also used in gene cloning. Some prokaryotic species benefit from other species but do not harm the other species. Such a relationship is called commensalism.

Commensalism: Commensalism is an interspecific interaction that positively (beneficially) affects one species and that is neutral toward the other. Some prokaryotic species establish this type of relationship with other species. For example, there are many bacterial species on our skin that nourish on the oils that seep through our pores. However these bacteria do not harm us. This is not to say that there are not any bacteria on our skin waiting for an opportunity to benefit by harming us, which brings us to parasites and pathogens.

Parasites and pathogens: A pathogen is any agent that infects a part of an organism such as an organ and causes disease. There are certain bacteria that act as pathogens. Infection is the process in which a pathogen bacterium multiplies inside the host organism and causes disease until it is halted. Diseases caused by bacteria kill millions of people each year.

Even though bacteria cause about half of all human diseases, other organisms such as viruses and some protists also cause diseases.

> Caution! All the known prokaryotic pathogens are bacteria, and not archaea; and are responsible for about half the human diseases. Yet, the large majority of bacteria (or prokaryotes in general) are useful and their role in survival of many species is crucial.

6.5 In a Nutshell

All life on our planet can scientifically be organized into three domains: Archaea, Bacteria, and Eukarya. Archaeans and bacteria are collectively called prokaryotes, which form the Monera Kingdom. Prokaryotes live almost everywhere, and outnumber and outmass all other life forms on Earth by a huge margin. The reproductive fitness and success of prokaryotes comes from their diversity that is displayed in their anatomical and physiological adaptations such as various metabolic adaptations. This diversity was facilitated by rapid reproduction followed by substantial mutations, and various genetic recombination methods.

Prokaryotic diversity can be appreciated by observing the wide range of relationships that prokaryotes have with the rest of life: beneficial to other species, harmful to other species, and neutral. Furthermore, prokaryotes play a crucial role in maintaining life in the biosphere by acting as decomposers and by establishing mutualistic relationships with other species.

Prokaryotes are unicellular and diverse, and outnumber eukaryotes. Within eukaryotes, however, there is a very diverse group of mostly unicellular organisms, called protists, which are the subject of the next chapter.

6.6 Review Questions

1. True or False: Most eukaryotes are multicellular.

2. True or False: Studies suggest that the flagella of archaea and bacteria evolved through convergent evolution.

3. What is true about Kingdom Monera?

   A. It consists of all the unicellular organisms.

   B. It consists of all the prokaryotes.

   C. It consists of bacteria and archaeans.

   D. B and C are true.

   E. A, B, and C are true.

4. Total biomass of all prokaryotes is:

   A. Much greater than the total biomass of all eukaryotes.

   B. Much smaller than the total biomass of all eukaryotes.

   C. About equal to the total biomass of all eukaryotes.

   D. Impossible to estimate for this comparison.

5. The physical transfer of a gene from an organism of one evolutionary lineage to another organism of a different lineage is called:

   A. Horizontal gene transfer

   B. Vertical gene transfer

   C. Lateral gene transfer

   D. Conjugation

   E. A, C, and D

6. Which of the following is not true about bacteria?

   A. They are prokaryotes.

   B. Most of them cause diseases.

   C. They give flavor to certain foods.

   D. They come in many shapes.

   E. They play important role in cycling nutrients.

7. Which of the following is true about bacteria?

   A. They aid in producing medicines.

   B. They cause diseases.

   C. Some of them are eukaryotes.

   D. They play important role in keeping the environment suitable for supporting life.

   E. A, B, and D are true

8. Bacteria reproduce through:

   A. Meiosis

   B. Mitosis

   C. Binary fission

   D. Lateral gene transfer

9. Prokaryotes are:

   A. Chemoheterotrophs

   B. Photoheterotrophs

   C. Chemoautotrophs

   D. Photoautotrophs

   E. All of the above

10. Which of the following has caused genetic variation in prokaryotes?

   A. Mutations

   B. Conjugation

   C. Transformation

   D. Transduction

   E. All of the above

6.7 Answer Key

1. False
2. True
3. D
4. A
5. E

6. B
7. E
8. C
9. E
10. E

Notes:

Q1. Most eukaryotes are unicellular.

Q2. Flagella in archaea and bacteria are analogous structures and not homologous.

Q3. It is true that most organisms in the Kingdom Monera are unicellular, but it does not mean that all unicellular organisms are in the Kingdom Monera.

Biology of Evolution and Systematics

Chapter 7

Evolution of the Protists: Collapse of a Kingdom

7.1 Protists: The Big Picture

In the five-kingdom classification scheme, all life on Earth is grouped into five kingdoms: Monera, Animalia, Plantae, Fungi, and Protista. We explored Kingdom Monera in Chapter 6, which consists of all prokaryotes and has been split into two domains: Archaea and Bacteria.

The Kingdom Protista (Figure 7.1), which once contained all eukaryotic microorganisms, has been found to be so diverse that the rank Kingdom Protista has collapsed. The former members of Kingdom Protista are no longer classified into just one kingdom. However, they are still called protists for sake of convenience. Here are some important facts about these protists:

1. **Eukaryotes:** They are eukaryotes occupying most of the lineages in the evolutionary tree of eukaryotic organisms.

2. **Mostly Unicellular:** They are mostly unicellular; some are multicellular.

3. **Polyphyletic:** They are so diverse that they form a polyphyletic group, not a monophyletic group, a clade.

Biology of Evolution and Systematics, by Paul Sanghera
Copyright © 2015 Infonential.

4. **Use organelles:** Because they lack organs, unicellular protists perform their essential functions by using their subcellular organelles such as the endoplasmic reticulum (ER), the Golgi apparatus, lycosomes, and the nucleus.

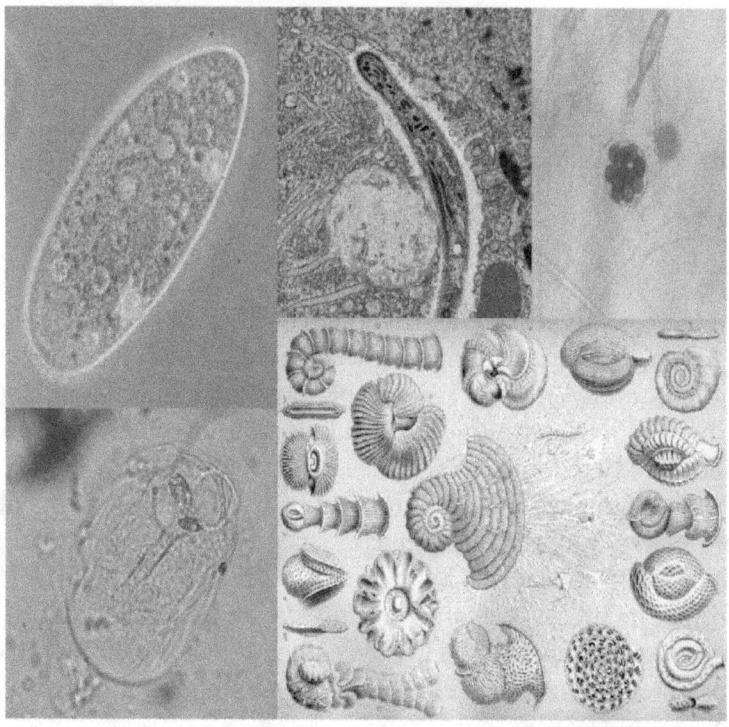

Figure 7.1 Kingdom Protista: Some examples of protists.

5. **Diverse modes of nutrition:** Some protists have chloroplasts and therefore are photoautotrophs. Some are heterotrophs that ingest food particles or directly absorb organic molecules. Others use both modes of nutrition: photoautotrophy and heterotrophy, and therefore are called mixotrophs.

6. **Diverse lifecycles:** Some protists produce sexually, some asexually, and others can use both modes, at least partially. Each of these three life cycles also vary among protists exhibiting features such as alternation of generation, lack of multicellular haploid, and lack of multicellular diploids.

7. **Endosymbiosis:** The diversity of protists mostly originated from a process called *endosymbiosis,* or *primary endosymbiosis,* in which some unicellular organisms engulf cells from their external environment. The engulfed cell is called an endosymbiont, and subsequently it turns into an organelle.

Caution! Over the years, what was known as the Kingdom Protista was found to be so diverse that it could not be organized into a clad, which means it is polyphyletic, that is not from a common ancestor.

Because unicellular organisms have been ingesting many different kinds of cells over time, the organisms took different paths of evolution, which gave rise to enormous diversity.

An example of endosymbiosis can be observed by studying the evolution of algae. There is enough DNA data to hypothesize that two photosynthetic lineages of algae, green algae and red algae, appeared because a eukaryote engulfed a cyanobacterium which evolved into plastids, a photosynthesis-related and color-related group of organelles such as chloroplasts, chromoplasts, and amyloplasts. DNA evidence also suggests that the heterotrophic eukaryotes that engulfed green algae and red algae gave rise to a group of protists called chlorarachniophytes, a small group of algae

119

found in tropical oceans. This process, in which a unicellular organism engulfs the product of a primary endosymbiosis, is called *secondary endosymbiosis*. Endosymbiosis is also discussed in Chapter 4.

7.2 Making Sense of Protists

The former Kingdom Protista served as a junkyard for biologists: when they found a species which did not really fit into the animal, plant, or fungi kingdom, they threw it into the Kingdom Protista. Therefore, it is no wonder that sometimes protists are grouped into three convenient categories: animal-like, plant-like, and fungus-like. This is how much commonality can be found within this very diverse group. Table 7.1 presents some common characteristics and examples of animal-like, plant-like, and fungus-like protists.

Table 7.1 Common characteristics and examples of animal-like, plant-like, and fungus-like protists.

Group	Varieties	Common characteristics and examples
Animal-like (Protozoa)	1. Protists with cilia 2. Protists with flagella 3. Protists with pseudopods 4. Parasitic protists	Unicellular, heterotrophs Most likely evolutionary ancestors of animals Live in fresh water, moist soil salt water, on or in other organisms Examples: amoeba, paramecium, plasmodium
Plant-like	1. Autotrophs 2. Mixotrophs: practice autotrophy and heterotrophy	Most are photosynthetic; produce oxygen Live on barks of trees, in fresh water, in salt water,

	3. Stripes: stem-like structures 4. Holdfasts: anchoring structures 5. Unicellular, colonial, multicellular, filamentous	and in soil Basis for the aquatic food chain Examples: euglenoids, dinoflagellates, chrysophytes, green, red, and brown algae
Fungus-like	1. Unicellular 2. Colonial 3. Multicellular	Heterotrophic Can act as decomposers. Live in moist environment, decaying products, soil, and trees Examples: slime molds and water molds

7.3 Evolution of Protists: Phylogenetics

The evolutionary relationships among all eukaryotes including protists revealed by the phylogenetic studies are shown in Figure 7.2. As this figure illustrates, eukaryotes are classified into five supergroups: Excavata, Chromalveolata, Rhizaria, Archaeplastida, and Unikonta. All the lineages in this tree, except animals, plants, and fungi, belong to protists. Here are the key characteristics of the five supergroups:

Excavata: This supergroup includes protists with unique flagella. These flagella are structurally different from the flagella found in other organisms. The group also includes protists with modified mitochondria.

Chromalveolata: DNA evidence indicates that this supergroup may have originated through secondary endosymbiosis, as their ancestors engulfed red algae.

Rhizaria: Molecular studies indicate that the supergroup Rhizaria is a monophyletic group, that is, a clade.

Archaeplastida: This group includes red algae and green

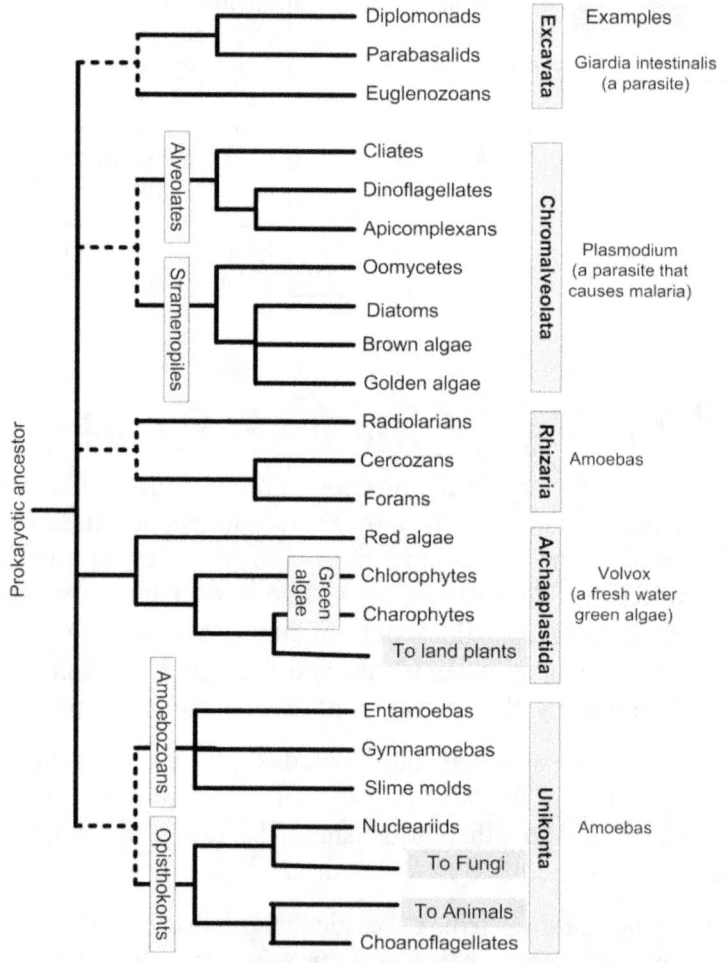

Figure 7.2 Evolutionary relationships among eukaryote supergroups and phyla. Dotted lines indicate the relationships that are still under study and debate. Animals, fungi, and land plants are not included in protists.

algae, which are the closest relatives of land plants.

Unikonta: This supergroup includes protists that are the closest relatives of fungi.

More characteristics and examples of the protists in these five supergroups are presented in Table 7.2 in the next section.

7.4 Examples and Characteristics of Protists

Table 7.2 presents some examples and characteristics of protists. Here are some notes about some entries in this table:

- Protozoa (meaning "first animals") are heterotrophic, single-celled or colonial eukaryotes.

- Chloroplasts in Euglena originated from green algae by secondary endosymbiosis.

- Chloroplasts in brown algae and diatoms originated from red algae by secondary endosymbiosis.

- Chloroplasts in dinoflagellates originated from red and green algae by secondary and tertiary endosymbiosis.

Fascinating Fact! Protists are so diverse that even the group known as Amoeba is not a clade, it is actually a polyphyletic group.

Biology of Evolution and Systematics

Table 7.2 Characteristics and examples of protists.

	Form	Motility	Nutrition	Functional Group
Excavata				
Euglena	Unicellular	Motile: Flagella	Mixotroph	Protozoa
Trypanosoma	Unicellular Ribbon-shaped	Motile: undulating membrane and flagellum	Parasite	Parasite: causes sleeping sickness in human
Chromalveolata: Alveolates				
Plasmodium (Apicomplexans)	Unicellular	Motile Lifestyle: intercellular	Parasitic Pathogenic	Parasite: causes malaria
Ciliates	Colonial Two types of nuclei: small and large	Motile: uses cilia to move and feed	Autotroph	Protozoa
Dinoflagellates	Unicellular Colonial	Motile: uses flagella to move and spin	Photosynthetic Heterotrophic Mixotrophic	Some are pathogenic: releases toxins Mostly marine organisms
Chromalveolata: Stramenopiles				
Diatoms	Mostly unicellular Some colonial Silica walls with holes	Mostly non-motile Some move with flagella	Autotroph Photosynthetic	Algae: freshwater and marine
Brown algae	Multicellular	Motile cells: flagella	Autotroph Photosynthetic	Algae Seaweeds: largest of the seaweeds
Rhizaria				
Forams	Unicellular porous shells made of calcium	Motile	Heterotroph A few are parasitic	Protozoa

	carbonate			
Radiolarians	Unicellular Internal skeleton made of silica	Non-motile Motile	Heterotroph	Protozoa
Arachaeplastid a				
Red algae (Rhodophyta)	Most are multicellular Cell walls contain agar	Non-motile cells	Autotroph: photosynthetic	Algae Seaweeds
Green algae: chlorophytes and charophytes	Unicellular Multicellular Filamentous Colonial Cell walls composed of cellulose	Mostly non-motile Some are motile: flagella	Autotroph: photosynthetic	Algae: mostly freshwater
Green algae: chlorophyta: spirogyra	Filamentus Spiral arrangement of chloroplasts	Non-motile	Autotroph: photosynthetic	Algae
Green algae: volvox	Colonial	Non-motile	Autotroph: photosynthetic	Algae
Green algae: ulva	Multicellular	Non-motile	Autotroph: photosynthetic	Algae
Unikonta				
Plasmodial slime molds	Unicellular Multinucleate	Motile	Heterotroph: phagophytic	Mold Fungus-like
Amoeba	Unicellular	Motile	Heterotroph Some are parasitic	Protozoa

7.5 In a Nutshell

All protists are eukaryotes, and they occupy most of the lineages in the evolutionary tree of eukaryotic organisms. Protists are very diverse and range from fungus-like to plant-like, to animal-like. Chloroplasts in some protists originated from green algae or red algae through secondary or tertiary endosymbiosis.

7.6 Review Questions

1. Charophyceans, brown algae, and red algae are some examples of protists that are:

   A. Parasitic

   B. Decomposers

   C. Photosynthetic

   D. Chemoautotrophic

   E. Photoautotrophic

2. Plasmodium, the parasite that causes malaria, is a:

   A. Unikonta

   B. Excavata

   C. Rhizaria

   D. Chromalveolata

3. A paramecium is a:

   A. multicellular organism

   B. protozoan

   C. metazoan

   D. bacteria

   E. dinoflagellate

4. When a protist engulfed a cyanobacterium, it evolved the chloroplasts of which organisms?

   A. Brown algae

   B. Green algae

C. Red algae

D. Both A and C

5. Which of the following protists are most closely related to Fungi?

A. Animals

B. Gymnamoebas

C. Green algae

D. Red algae

6. Which of the following is not a true statement:

A. Kingdom Protista is not monophyletic.

B. Amoebas are not monophyletic.

C. Protists use organs to perform functions.

D. Excavates include protists with modified mitochondria.

7. Sleeping sickness is caused by:

A. Trypanosoma

B. Euglena

C. Amoeba

D. Plasmodium

8. The chloroplasts of Euglena are evolved from green algae through:

A. Primary Endosymbiosis

B. Secondary Endosymbiosis

C. Horizontal gene transfer

D. Vertical gene transfer

9. Which of the following is the largest of the seaweeds?

A. Brown algae

B. Red algae

C. Green algae

D. None of the above

10. Which of the following are not all protists?

A. Excavata

B. Chromalveolata

C. Rhizaria

D. Archaeplastida

7.7 Answer Key

1.	C	6.	C
2.	D	7.	A
3.	B	8.	B
4.	D	9.	A
5.	B	10.	D

Notes:

Q5. Animals are not protists.

Q6. Protists use organelles, and not organs, to perform functions.

Q10. Archaeplastida is a supergroup of eukaryotes that consists of protists and land plants.

Biology of Evolution and Systematics

Chapter 8

Evolution of the Kingdom Plantae

8.1 Kingdom Plantae: The Big Picture

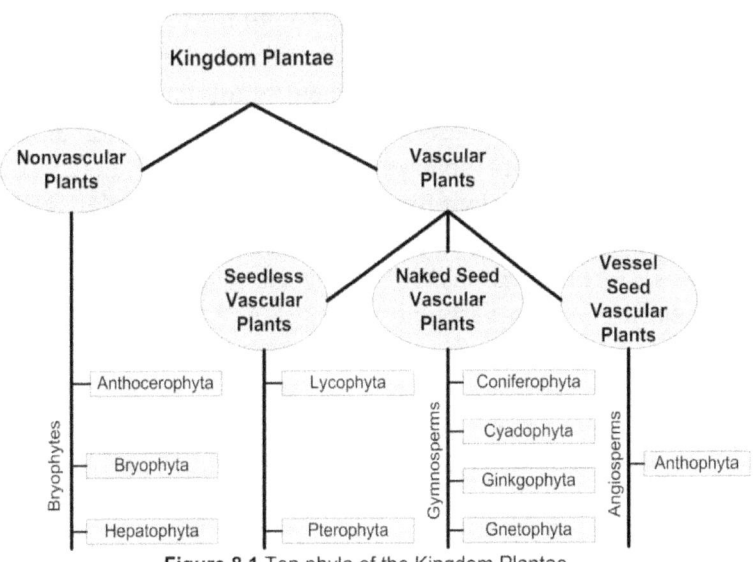

Figure 8.1 Ten phyla of the Kingdom Plantae

The first land plants appeared about 475 million years ago, during the Paleozoic era. As shown in Figure 8.1, the

Biology of Evolution and Systematics, by Paul Sanghera
Copyright © 2015 Infonential.

Biology of Evolution and Systematics

Kingdom Plantae consists of two types of plants: vascular and non-vascular. Vascular plants are plants that have vascular tissues, a phloem and a xylem. The vascular tissue system helps a plant to remain standing upright, and it transports and distributes carbohydrates, mineral, and water within the body of the plant. The xylem of almost all the vascular plants contains tube-shaped cells called tracheids, which carry fluid (water and minerals) from the roots to the upper parts of the plant. Plants are multicellular eukaryotic organisms that make their own food using light from the sun, carbon dioxide from the atmosphere, and water from the ground by using a process called photosynthesis. For this reason, they are called autotrophs, photoautotrophs to be exact.

The vascular tissue system combined with certain reproductive traits gives a competitive edge to vascular plants, so much so that ninety percent of all plant species and seven out of the ten plant phyla are vascular. The remaining ten percent of plant species do not have well developed vascular tissue and therefore are called non-vascular. Non-vascular plants are grouped into three phyla: Bryophyta (mosses), Anthocerophyta (hornworts), and Hepatophyta (liverworts). Non-vascular plants are collectively called bryophytes.

Note. There are more than 290,000 known species of plants living on our planet today, which are grouped into ten different phyla.

Vascular plants are further classified by those that produce seeds and those that do not.

Seedless vascular plants are grouped into two phyla: Lycophyta, collectively called lycophytes (club mosses, spike mosses, quillworts), and Pterophyta collectively called pterophytes (ferns, whisk ferns, horsetails).

There are two types of seed vascular plants: naked-seed vascular plants and vessel-seed vascular plants. The naked seed vascular plants are the plants that do not have their seeds enclosed, such as pine cones. The vessel-seed vascular plants are the plants whose seeds are enclosed in vessels, such as fruits. Naked seed plants, collectively called gymnosperms, are grouped into four phyla: Coniferophyta (conifers), Cyadophyta (cyads), Gingophyta (ginkgos), and Gnetophyta (Gnetophytes). The vessel-seed plants, collectively called angiosperms, are grouped into only one phylum: Anthophyta (flowering plants).

Table 8.1 Brief information about the Kingdom Plantae.

Plant Type	Phylum	Common Name	Number of Known Species
Nonvascular Plants	Bryophyta	Mosses	100
	Anthocerophyta	Hornworts	15,000
	Hepatophyta	Liverworts	9,000
Seedless Vascular Pants	Lycophyta	Lycophytes (club mosses, spike mosses, quillworts)	1,200
	Pterophyta	Pterophytes (ferns, whisk ferns, horsetails)	12,000
Naked Seed Vascular Plants (Gymnosperms)	Coniferophyta	Conifers	600
	Cycadophyta	Cycads	130
	Ginkgophyta	Ginkgos	1
	Gnetophyta	Gnetophytes	75
Vessel-Seed Vascular Plants (Angiosperms)	Anthophyta	Flowering Plants	250,000

The information about all these plant phyla is summarized in Table 8.1.

Fascinating Fact! The vessel seed vascular plants, angiosperms or flowering plants, belong to only one phylum, Anthophyta, but consist of almost ninety percent of all plant species.

8.2 Reproductive System in Plants

The life cycle of all plants is determined by a process called *alternation of generations* in which two alternate generations give rise to each other. These two alternate generations are created from a multicellular haploid form called *gametophyte* and a multicellular diploid form called *sporophyte*. The steps of this cycle are illustrated in Figure 8.2. In non-vascular plants, the gametophyte stage is dominant. For example, when you look at a green moss growing on a rock, you are looking at a gametophyte. As illustrated in Figure 8.2, the gametophyte produces gametes (sperms and eggs) by mitosis. The sperm and the egg fuse through fertilization to form a zygote, which subsequently produces spores through meiosis. These spores can develop into sporophytes through mitosis. Once plants evolved to live on land, natural selection favored sporophytes, which adapted well with the drier environment. This is how sporophyte dominance developed. For example, pine trees, which are native to most of the Northern Hemisphere and are diversified into about 115 species, are sporophytes.

Table 8.2 summarizes some reproductive features of plants. It clearly shows the continuous change from gametophyte dominance to sporophyte dominance, homosporous feature to heterosporous feature, and unisexuality to bisexuality during the evolutionary journey from non-vascular plants to vessel-seed vascular plants.

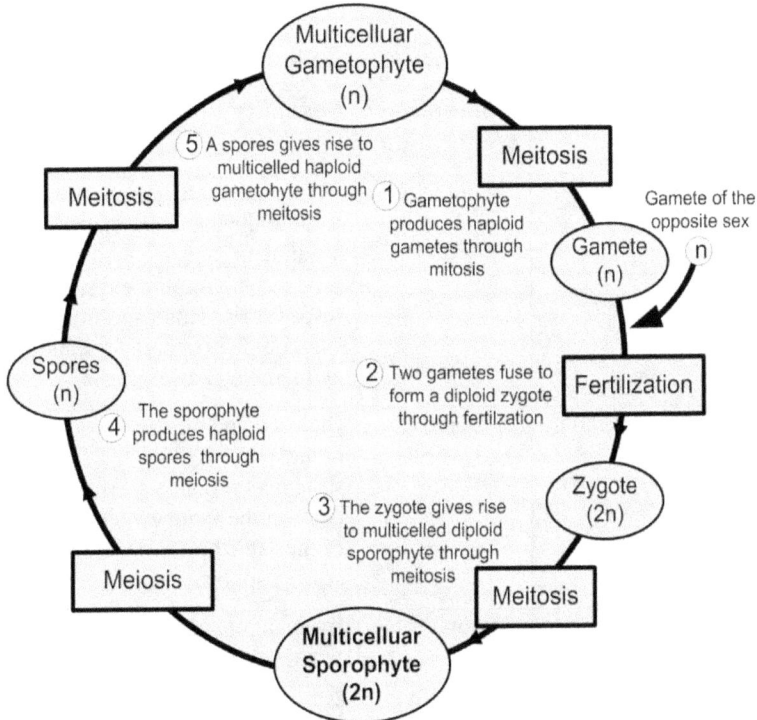

Figure 8.2 Generalized lifecycle of plants.

Caution! Do not confuse spores with gametes. Spores, although they do develop into gametophytes, have nothing to do with fertilization directly.

Table 8.2 Reproductive features of plants.

Group	Dominant Generation	Sexuality
Nonvascular Plants	Gametophyte	Unisexual: male gametophytes and female gametophytes are separate organisms, which produce sperms and eggs respectively. Homosporous: single type of spores for male and female gametophytes.
Seedless Vascular Plants	Sporophyte	Bisexual and mostly homosporous. Male and female structures produce sperms and eggs on the same gametophyte of the same organism. Single type of spores.
Naked seed Vascular Plants (gymnosperms)	Sporophyte	Bisexual plants. Male and female structures produce sperms and eggs via different gametophytes within the same organism. Male and female cones on the same plant. Heterosporous: two types of spores; microspores that develop into male gametophyte and megaspores that develop into female gametophyte.
Vessel Seed Vascular Plants (Angiosperms or flowering plants)	Sporophyte	Bisexual. Male and female structures produce sperms and eggs via different gametophytes within the same organism. Most

		plants have male and female structures in the same flower. Heterosporous: microspores and megaspores.

Note these three critical points related to all plants:

- All plants have an alternation of generation lifecycle.

- Meiosis produces haploid spores, not gametes (sperms and eggs). A spore produced by a sporophyte develops into a gametophyte. A structure called *sporangium* in a sporophyte produces spores.

- A gametophyte produces gametes through mitosis, and not meiosis. A structure called, *gametangium* in a gametophyte produces gametes. A male gametangium, called *antheridium* produces sperms, and a female gametangium, called *archegonium*, produces eggs.

For all seedless plants, note the following:

- The sperms must swim through an external environmental water to reach the egg. This is why an external watery environment is necessary for the sexual reproduction of seedless plants.

- Plants are dispersed to new locations as spores, and not as seeds.

8.3 Plants: On the Path of Evolution

It is evident that land plants evolved from algal ancestors. The closet relative of plants on the evolutionary tree, as scientists have determined from the molecular data, is a form of green algae called charophytes. Charophytes and plants share many similar characteristics, such as chloroplasts. Not only do both have chloroplasts, but their chloroplasts have the same chlorophyll molecules, named a and b. The cladogram illustrating the evolution of plants is presented in Figure 8.3.

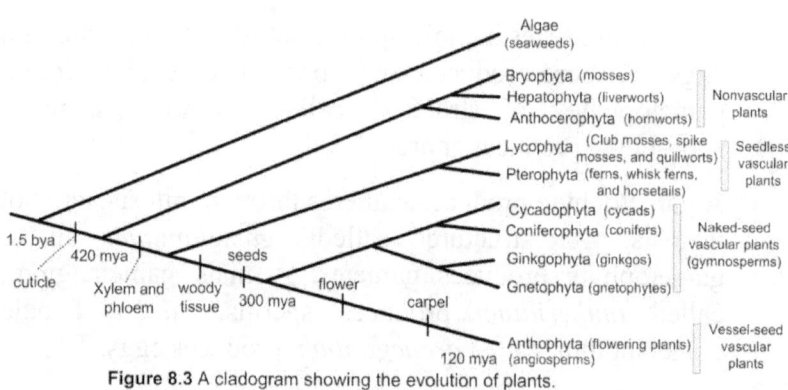

Figure 8.3 A cladogram showing the evolution of plants.

Many species of algae are known to do well living on the interface of water and land, such as bodies of shallow waters around the perimeters of lakes and ponds. They can survive occasional drying to an extent. So, the origin of the Kingdom Plantae can be thought of as a spin out onto the land by algae sitting on the interface of water and land. From the very beginning of their journey of evolution to the survival and reproduction in current times, plants went through three major evolutionary pressures, under which the anatomy and physiology of their reproductive and other systems evolved. These three pressures, defined by the condition of living and reproducing on land (with scarcity of water as compared to the sea) without moving, are the following:

1. **Transportation:** How to transport sperms to eggs for sexual reproduction? In general, how to disperse propagules (the parts of plants that facilitate sexual or asexual reproduction)?

2. **Protection:** How to protect the entities of the reproduction process (gametes, zygote, and developing embryo) from dehydration?

3. **Resources:** How to get resources (they need water, minerals, carbon dioxide, and light) from two different parts of their environment from below the ground surface, and from above the ground surface?

The fossil record accounts for the four major phases of plant evolution: the origin of land plants about 475 mya, the appearance of early seedless vascular plants around 420 mya, the appearance of early naked seed vascular plants around 360 mya, and the radiation of flowering plants around 140 mya. The story of the evolution of plants is a journey of increasing adaptation to life on land, and thus to relatively dry environment.

> Note. Gymnosperms, the cone bearing plants, were dominant along with dinosaurs in the Mesozoic era, whereas the Cenozoic era is the era of the angiosperms, the flowering plants.

Table 8.3 presents some of the traits that have evolved in plants. The digits 0, 1, 2, and 3 that are used in this table for traits have the following meanings:

- **Embryo, seeds, vascular tissue, flowers :**

  $0 \equiv$ absent $1 \equiv$ present

- **Leaves (on sporophyte):**

0 ≡ absent 1 ≡ microphylls 2 ≡ megaphylls

- **Life cycle:**

  0 ≡ no alternation of generations

  1 ≡ alternation of generations with sporophyte depending on gametophyte for nutrition

  2 ≡ alternation of generations with sporophyte independent of gametophyte; both free-living

  3 ≡ alternation of generations with gametophyte depending on sporophyte for nutrition

Table 8.3 Evolution of traits in plants.

	Embryo	Leaves	Seeds	Vascular tissue	Flowers	Life cycle
Charophytes	0	0	0	0	0	0
Mosses	1	0	0	0	0	1
Lycophytes	1	1	0	1	0	2
Ferns	1	2	0	1	0	2
Gymnosperms	1	2	1	1	0	3
Angiosperms	1	2	1	1	1	3

Problem 8.1 Draw a cladogram from the data in Table 8.3.

Solution:

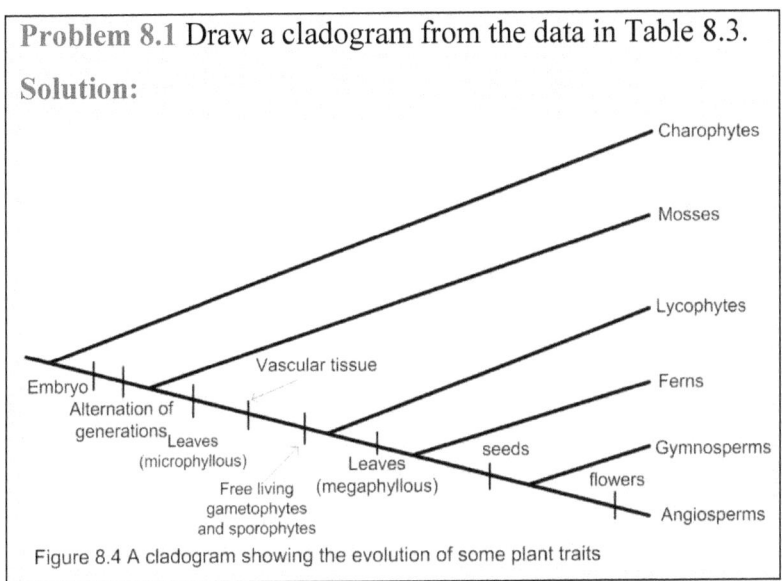

Figure 8.4 A cladogram showing the evolution of some plant traits

8.4 Key Evolutionary Adaptations in Plants

Following are the key adaptations, the key derived traits, that are present in all plants collectively called embryophytes, but not in their ancestor algae and their closest algae relative the charophytes. The plants owe their name embryophytes to the embryo, one of their four key adaptations.

1. **Multicellular dependent embryo:** The alternation of generations, already discussed in section 8.2, refers to the life cycle of land plants, which alternates between two generations of multicellular organisms, gametophytes and sporophytes, generating each other. Although the process evolved in various groups of algae, it does not occur in the algae called charophytes, the closest relative of plants. An important part of the alternation of generations life cycle is the multicellular plant embryo, which develops from the zygote. The embryo is dependent on its female parent, the gametophyte, which retains it within its tissues that nourish it with nutrients such as amino acids and sugars.
   Make sure you understand that alternation of generations.

2. **Multicellular gametangia:** Another adaptation that distinguishes the early plants form their algal ancestors is an organ called the *gametangia*, which produces gametes. The female gametangia is called an *archegonium* (plural: *archegoina*) and produces a single nonmotile egg. The male gametangia is called an *antheridium* (plural: *antheridia*) and it produces sperms.

3. **Multicellular sporangia:** During the lifecycle of a plant, a multicellular organ called the *sporangium* (plural: *sporangia*) in a sporophyte produces haploid cells called spores by meiosis. A spore can develop into a multicellular gametophyte by mitosis. The walls of these spores are tough enough to survive through most harsh environments, because they are made of a polymer called sporopollenin. Even though charophytes also produce

spores, they do not have sporangia, and their spores lack sporopollenin walls.

4. Apical meristems: An epical meristem is an embryonic plant tissue (with undifferentiated cells) found in the buds and in the growing tips of roots. The undifferentiated cells in the meristem can divide rapidly and a fraction of the descendent cells then differentiate into specialized tissues. This helps the plant to grow in length.

Caution! Make sure you do not confuse the alternation of generations in plants with the haploid and diploid stages during the life cycle of other sexually reproducing organisms, such as human. In humans, even though two haploid gametes (from a male and a female) fuse to form zygote, this haploid stage, unlike in plants, does not involve the multicellular haploid organism called gametophyte.

Here is a word of warning. Do not assume that all of these traits are unique to plants and that all plants carry these traits. Some of these traits have developed independently in other (non-plant) evolutionary lineages (think of convergent evolution), and some of these traits, such as gametangia, have been lost over time in some plant lineages.

Caution! Tracheids are found in all vascular plants, whereas vessels are only found in angiosperms and one phylum of gymnosperms, the gnetophyta.

A great majority of plants these days are vascular seed plants. In addition to the key adaptations discussed above, some other adaptations that helped plants to colonize land include cell walls, chloroplast, cuticle, lignin, stomata, vascular tissue, pollen, seed with protected embryo, roots and shoots, and mycorrhizae. Two of these characteristics, cell walls and chloroplast, are shared with algae.

8.5 More definitions of Terms and Concepts

Sporophyll: A modified leaf that bears sporangia and is therefore a reproductive part of a plant.

Sporangium: An organ in plants in which meiosis occurs to produce haploid cells called spores.

Microphylls: Small and mostly spine-shaped leaves that only have one vein, which is unbranched. Lycophytes and only lycophytes (club mosses, spike mosses, quillworts) have such leaves.

Megaphylls: Large leaves that have a network of veins. Ferns and seed vascular plants have such leaves.

Rhizoid: A tube-shaped (hair-like) cell or a filament of a cell. Rhizoids make the root-like structure, which anchors the gametophytes of bryophytes (fern, mosses, and liverworts) to the ground.

Roots: Similar to how rhizoids evolved in bryophytes (nonvascular plants), roots evolved in almost all vascular plants. Roots absorb fluid (water and nutrients) from the soil and anchor the plant to the ground. In this way roots help the plant to grow taller.

Leaves: Leaves are the primary photosynthetic organs of vascular plants.

Rhizomes: Horizontal stems. The leaves of most ferns grow from rhizomes.

Sorus (plural: *sori*): A little brown dot under the leaf of a fern. This structure contains dozens of sporangia.

Strobilous (plural: *strobili*): A cone-like structure, formed of sporophylls. Found mostly in gymnosperms, but also in many lyclophytes.

Monocot: Any flowering plant that has one cotyledon, the embryonic seed leaf.

Dicot: Any flowering plant that has two cotyledons, the embryonic seed leaves.

Eudicot: Any flowering plant that belongs to a subset of dicots that make a clade.

Angiosperms (flowering plants) fall into two categories: monocots and dicots, which are compared in Table 8.4.

Table 8.4 Comparing monocots and dicots.

Characteristic	Monocots	Dicots
Cotyledons (seed leaves)	One	Two
Leaf venation	Usually parallel	Net-like
Roots	Fibrous	Taproot (primary root) present
Vascular tissues in stems	Scattered	Arranged in a ring
Floral parts in	Multiples of three	Multiples of four or five

Pollen grain with	One opening	Three openings

8.6 In a Nutshell

Plants are photoautotrophs. Molecular studies suggest that the ancestors of plants were similar to green algae, as charophytes, a form of green algae, are the closest relatives to plants. Plants fall into two main categories: vascular plants that have specialized tissues called xylem and phloem to transport fluid, and non-vascular plants, which do not have these tissues. Vascular plants fall into two main categories: seedless vascular plants and seed vascular plants. Pollen, grains, and seeds are among the key adaptations of plants that enabled them to live on land. The key reproductive adaptations of angiosperms include flowers and fruits.

The lifecycle of all plants is determined by a process called alternation of generations, meaning the lifecycle alternates between two generations: gametophytes and sporophytes. The lifecycle of nonvascular plants is dominated by gametophytes, whereas the lifecycle of vascular plant is dominated by sporophytes.

8.7 Review Questions

1. Flowering plants are most closely related to:

   A. Angiosperms

   B. Gymnosperms

   C. Lycophyta

   D. Bryophyta

2. The extant plants that are most similar to the first plants that had gametangia are:

   A. Charophytes

   B. Lycophytes

   C. Bryophytes

   D. Anthophytes

3. Which of these characters the plants share with algae?

   A. Roots

   B. Shoots

   C. Vascular tissue

   D. Chloroplast

   E. Pollen

4. In a moss, which of the following produces sperms?

   A. Sporophyte

   B. Sporangium

   C. Archegonium

   D. Antheridium

5. When you are looking at a fern plant from a distance, what are you looking at?

   A. A haploid sporophyte

   B. A diploid sporophyte

   C. A haploid gametophyte

   D. A diploid gametophyte

Biology of Evolution and Systematics

6. What was the age of gymnosperms?

   A. Mesozoic era

   B. Paleozoic era

   C. Proterozoic eon

   D. Cenozoic era

7. Which of the following is not a correct statement about the lifecycle of plants?

   A. Meiosis produces gametes.

   B. Antheridium produces sperms.

   C. Gametangium produces eggs.

   D. A spore develops into a gametiphyte.

8. Which of the following is not true about ferns?

   A. These are seedless vascular plants.

   B. These have microphyllous leaves.

   C. In their lifecycle, both gametophytes and sporophytes are free-living.

   D. They have embryos.

9. The ancestors of plants were the most similar to which of the following existing groups of organisms?

   A. Mosses

   B. Ferns

   C. Green algae

   D. Fungus

10. Which of the following are not vascular plants?

A. Hornworts

B. Club mosses

C. Connifers

D. A and B

8.8 Answer Key

1.	B	6.	A
2.	C	7.	A
3.	D	8.	B
4.	D	9.	C
5.	B	10.	A

Notes:

Q7. Meiosis produces haploid spores (and not gametes). A gametophyte produces gametes through mitosis.

Q8. Ferns have megaphyllous leaves.

Q10. Club mosses are seedless vascular plants.

Chapter 9

Evolution of the Kingdom Fungi

9.1 Kingdom Fungi: The Big Picture

Unity behind diversity applies better to the Fungi Kingdom than to the Protista Kingdom. Unlike Kingdom Protista, the members of Kingdom Fungi are more clearly defined and have closer evolutionary relationships among themselves. Fungi make a clade, whereas protists do not. Some examples of fungi are presented in Figure 9.1.

Almost all of the members of Kingdom Fungi have the following characteristics:

Figure 9.1 Examples of fungi: a) *Amnita muscaria*, a basidiomycete; b) *Sarcoscypha coccinea*, an ascomycete; c) a *Penicillium conidiophore;* d) a chytrid; and e) mold covering bread.

- They are mostly multicellular with a few exceptions such as yeast, which is unicellular.

Biology of Evolution and Systematics, by Paul Sanghera
Copyright © 2015 Infonential.

- They are eukaryotes, belonging to the domain Eukarya.

- Their cell walls are made of chitin, a modified polysaccharide.

- They are heterotrophic, like animals. However, unlike animals, fungi are heterotrophic by absorption, that is, they obtain their nutrition by absorption instead of ingesting. Organisms that do this are called *decomposers*.

- They have a haploid-dominant life cycle.

- During their lifecycle, fungi exhibit a stage called the *heterokaryotic stage*, in which there are two or more haploid nuclei per cell; these nuclei are contributed to by two different parents. This stage results from the fusion of cytoplasms from two parent *mycelia* (defined in section 9.2), in a process called *plasmogamy*.

- Their cells have no centrioles, a subcellular structure in animal cells, and the DNA in their chromosomes has only a few histones associated with it.

- With the exception of the phylum Chytridiomycota, fungi have no cilia or flagella at any stage of their life cycle.

- Mitosis during cell division takes place in an oddly interesting way: chromosomes divide within the nucleus, but the nuclear envelope does not break down. This is sometimes called *closed* mitosis.

Example of a Unicellular Fungus: A yeast is a unicellular fungus, and it uses two methods for reproduction:

- Binary fission.
- Budding; the pinching of a small bud off a parent cell, which will grow into the complete organism (cell in this case).

The Kingdom Fungi consists of over 100,000 species, which are assigned to five major groups (also called phyla): the Ascomycota, the Basidiomycota, the Chytridiomycota, the Glomeromycota, and the Zygomycota. This division is largely based on the kinds of structures fungi use for sexual reproduction.

> Form fits the function! The basic fungal body form is filamentous, which is elegantly adapted to facilitate its specific lifestyle: heterotrophic by absorption, or decomposer.

9.2 Structural and Functional Traits of Fungi

Here are some important structural and functional components and traits of fungi:

Hyphae: When you look at a fungus such as a mushroom, you are looking at a filamentous body form, which is the basic body form of fungi. The basic units of this form, the filaments, are called hyphae.

Mycelium: A mycelium is the whole collection or network of hyphae in a fungus. A fungus uses its mycelium to absorb or permeate food material.

Septa: These are the cross-walls that separate fungi cells from one another. The existence and size of the pores in these walls vary across phyla, resulting in diversity. For example, in the phylum Ascomycota, the pores in the septa are large enough that they allow the migration of a cell nucleus from one cell to another, yet in the phylum Basidiomycota, the pores are so small that they do not allow the passage of any nuclei, and in the phylum Zygomycota, sepata does not exist at all.

Exoenzymes: The digestive enzymes that fungi secrete out to their external environment in order to digest the food before they absorb it are called exoenzymes. This adaptation is what makes fungi the decomposers. It evolved because temperature and chemical conditions are more variable outside than inside the digestive tract of fungi, and therefore exoenzymes are needed for digestion.

Mycorrhizae: Mycorrhizae are the roots of a fungus and represent a mutualistic, symbiotic relationship between a host plant and the fungi that colonizes the host plant roots. The fungal hyphae help the plant's roots to increase the efficiency of water and nutrient absorption from underneath the surface, and in return the plant helps the fungi by providing it with photosynthetic products such as carbohydrates.

The type of mycorrrhizae relationship divides fungi into two categories: ectomycorrhizal fungi and endomycorrhizal fungi.

Ectomycorrhizal fungi: This type of fungi has hyphae that cover the intercellular spaces of the host organism's roots, but do not penetrate into the cells.

Endomycorrhizal fungi: These fungi are also called arbuscular mycorrhizal fungi. Their hyphae penetrate the root cells of the host organism.

Fascinating Fact! Almost all vascular plants have a mycorrhizal relationship with fungi in order to retrieve needed nutrients from the ground.

Lichen: This structure represents a mutualistic symbiotic relationship between a fungus and a photosynthetic partner, which is usually a green alga or a cyanobacterium. A lichen forms when a cell on the tip of a fungal hypha binds to a suitable photosynthetic cell, then after losing their cell walls, both cells divide, which results in the formation of a multicelled body, a lichen. Depending on many factors, lichen may have any of multiple shapes: erect, flattened, leaf-like, or pendulous.

Examples of lichens and mycorrhizal fungus are presented in Figure 9.2.

a)   b)

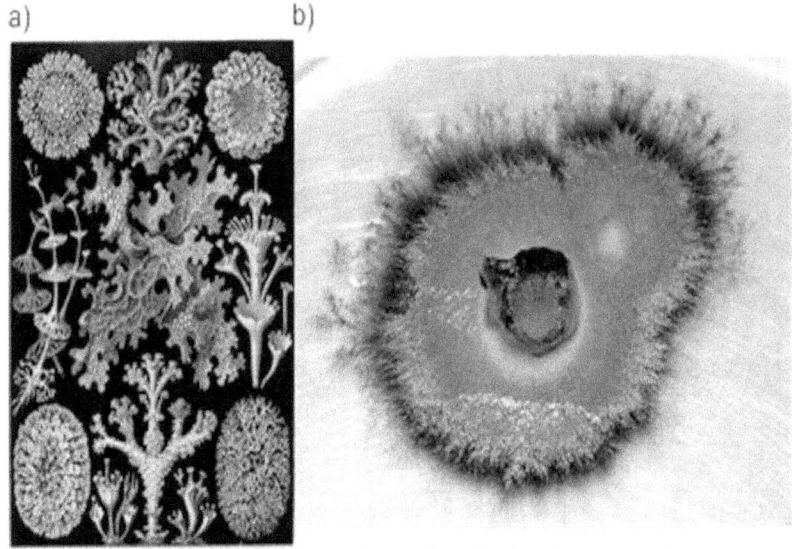

Figure 9.2 a) Depiction of organisms classified as lichenes. Courtesy: Ernst Haeckel's *Kunstformen der Natur*, 1904. b) An ericoid mycorrhizal fungus isolated from *Woollsia pungens* by David Midgley.

Biology of Evolution and Systematics

Table 9.1 summarizes some fungi-related terms, concepts, and traits.

Table 9.1 Some fungi-related terms.

Term	Description
Ascocarp	fruiting body of Phylum Ascomycota; often cup-shaped
Ascus	sac-like reproductive structure characteristic of Phylum Ascomycota; where meiosis occurs to produce eight ascospores (meiosis is followed by mitosis)
Basidiocarp	fruiting body of Phylum Basidiomycota; often mushroom-shaped
Basidium	club-like reproductive structure characteristic of Phylum Basidiomycota, where meiosis occurs to produce four basidiospores
Chitin	characteristic component of fungal cell walls
Conidia	asexual spores, produced by mitosis
Coenocytic	hyphae without septa as in the phylum Zygomycota
Haustoria	specialized hyphae in some fungal species used to exchange nutrients with host plants, for example, in a mutualistic relationship
Heterokaryotic	cells containing two distinct types of haploid nuclei
Hyphae	a filament, the basic unit of fungi's filamentous form
Karyogamy	fusion of nuclei
Lichen	mutualistic relationship between a fungus and a photosynthetic partner usually a green alga or a cyanobacterium
Mycelium	mass or network of hyphae in a fungus
Mycologist	a biologist who studies fungi
Plasmogamy	fusion of cytoplasm
Zygosporangium or Zygospore	reproductive structure characteristic of Phylum Zygomycota

> **Caution!** There are three kinds of fungus species: those that can grow only as unicellular yeast, those that can grow only as multicellular filaments, and those that can grow as both yeast and filaments.

9.3 Evolution of Fungi

As indicated in Chapter 7, fungi and animals are more closely related to each other than to any other group of eukaryotes including plants. This claim is supported by molecular data and fossil records.

The oldest fossils of fungi date back to about 460 mya and are from from the Paleozoic era. Data suggests that the ancestor of fungi was a unicellular, flagellated protist, most likely an amoeba.

The five phyla of fungi, as shown in the cladogram in Figure 9.3, are Chytridiomycota, Zygomycota, Glomeromycota, Ascomycota, and Basidiomycota.

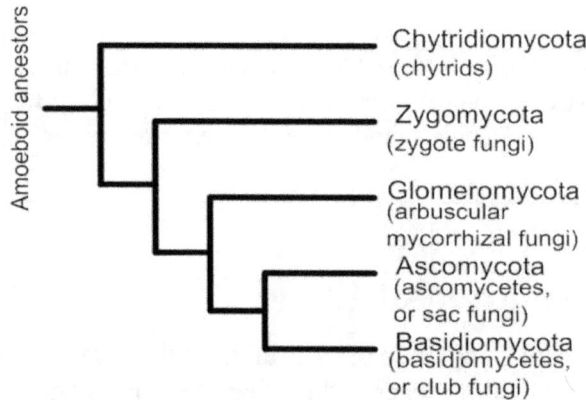

Figure 9.3 A Cladogram of the five phyla of fungi. However, chitrids and zygomycetes are not yet considered monophyletic groups.

Chytridiodimycota: According to molecular data, the chytrid lineage diverged early in the evolutionary tree. Based on up-to-date data, however, chytrids form a grade, not a clade, in the phylogenic tree. Unlike all other groups of fungi, the chytrids have flagellated spores called *zoospores*. They are found in water or in moist environments such as damp soil. Some chytrid species are decomposers, some are parasitic, and some are those that establish mutualistic symbiotic relationship with other organisms. For example, in 1998 the chytrid *Batrachochytrium dendrobatidis* was discovered to cause the disease in amphibians, called chytridiomycosis. On the other end of the spectrum, there are anaerobic chytrids that live inside the digestive tracts of the cattle, helping to break down and digest plant material.

Zygomycota: As discussed in section 9.2, unlike the hyphae of ascomycota and basidiomycota, hyphae of zygomycota fungi (zygotes) are coenocytic, that is, there are no septa dividing the hyphae into distinct cells. The distinctive feature of this group is the structure called the zygosporangium, a metabolically inactive sexual entity, resistant to freezing and drying. Zygosporangium is a multinucleate reproduction structure that zygotes make during reproduction, and it is where the processes of karyogamy and meiosis occur. Karyogamy is the fusion of two haploid cell nuclei by two parents, but it is different from the fertilization that produces a zygote. Whereas a zygote is a cell with one diploid nucleus, a zygosporangium has two states: one before karyogamy occurs in which it has many haploid nuclei from two parents, and the other after karyogamy occurs in which it has many diploid nuclei. Chances are that you have seen the most familiar representative of the phyla zygomycota: rhizopus, the common bread mold.

Glomeromycota: The distinctive feature of glomeromycota is that almost all glomeromycetes form symbiotic relationships with plants in the form of endomycorrhizae. The cells on the tip of the hyphae, push into the host plant's roots to form a tree-like branching pattern within the plant's root cells. Due to this tree-like branching, glomeromycetes are also known as *arbuscular mycorrhizae*.

Ascomycota: With 65,000 identified species, ascomycetes, also called cup fungi or sac fungi, are the most diverse of all fungi phyla in terms of the number of species. This phylum drives its name from its distictive feature, the asci, a saclike structures wherein sexual spores called ascospores are produced. Ascomycetes reproduce asexually by producing an enormous number of asexual spores called conidia. Examples of ascomycetes include yeast, which is unicellular, and cup fungi, which is multicellular. Another example of ascomycetes is the fungus penicillium used to produce antibiotics such as penicillin. Ascomycota fungi also vary in their functionality, ranging from being pathogens to being mutualistic symbiots. Lichens for example, are ascomycetes establishing a mutualistic symbiotic relationship with photosynthetic partners. Other ascomycetes are mycorrhizal with plants.

Basidiomycota: If you have seen a mushroom, you are already introduced to the phylum Basidiomycota. This is the most familiar phylum with members such as mushrooms, puffballs, and bracket (or shelf) fungi. The distinguishing feature of the basidiomycetes is that they reproduce sexually by producing elaborated fruiting bodies called basidiocarps, which produce sexual spores called basidiospores. The white mushroom that you use as food is a good example of a basidiocarp. The basidiomycetes play an important ecological role in the world, because they are the only fungi with the ability to breakdown lignin, the compound that stiffens plant stems. The decomposition of fallen forest plants is due to basidiomycetes.

Fascinating Fact! The fungal growth rate is so high that a single mycelium has the capability to grow 1 kilometer of new hyphae every day. Fungi use this capability to access distantly located new resources by expanding because they cannot move like animals can.

Table 9.2 presents the number of species and distinguished characteristics of the five phyla of fungi.

Table 9.2 The number of species and distinguishing features of the fungi phyla.

Phyla	Number of Species	Distinguishing Features
Chytridiomycota (chytrids)	1,000	flagellated spores called zoospores
Zygomycota (zygote fungi)	1,000	zygosporangium, a metabolically inactive sexual stage, resistant to freezing and drying
Glomeromycota (arbuscular mycorrhizal fungi)	160	forms arbuscular mycorrhizae with plants
Ascomycota (ascomycetes or sac fungi)	65,000	produces a enormous number of asexual spores called conidia; also produces sexual spores called ascospores in saclike structures called *asci*
Basidiomycota (basidiomycetes or club fungi)	30,000	reproduces sexually by producing elaborated fruiting bodies called basidiocarps, which produce sexual spores called basidiospores

Caution! Unlike in animals and plants, zygotes in fungus do not go through mitotic division. Instead they go through meiosis giving rise to the haploid dominant lifecycle.

Even though some of them are pathogens, fungi have beneficial practical uses, natural and artificial.

9.4 Pearls and Perils of Fungi

Fungi play an important role in ecology because of their mutualistic symbiotic relationship with other organisms. By decomposing (breaking down) the bodies and parts of dead organisms, they play important role in nutrient cycling, that is, the recycling of chemical elements between the living and the non-living. Some fungi protect plants from herbivores, animals that eat plants. Conversely, some fungi help animals to digest the plants that they eat. Humans use fungi for various applications, such as for food, in manufacturing products such as antibiotic penicillin and the yeast to brew beer and wine, and in several areas such as agriculture and forestry.

However, about thirty percent of all known fungi species are parasites, mostly infecting plants. There are also pathogenic fungi, which cause diseases in some groups of organisms such as animals and plants.

Think About It!

Q. What would happen to life on Earth in absence of fungi?

A. Nutrient cycling (or recycling) is critical for the continuation of life. If there were no fungi, there would be no breaking down of dead bodies, and therefore the nutrients essential for life would remain tied up in the dead organic matter. This would halt the nutrient cycling and as a result the life would ultimately cease to exist.

9.5 In a Nutshell

Molecular data suggests that fungi evolved from a flagellated and unicellular protist ancestor. Fungi and animals are close evolutionary relatives, as they have evolved (diverged) from the same ancestor. All fungi acquire their food or energy from chemicals such as organic compounds, which they break down if necessary, before absorbing them. This makes fungi chemoheterotrophs. Fungi have haploid dominant lifecycle. Some species reproduce asexually, others sexually. Some fungi species live as decomposers and others as symbionts. Whereas fungi play critical role in nutrient cycling, thereby maintaining life on earth, some of them are parasitic and pathogenic.

Caution! Although mushrooms are used as food, the difference in the morphology (appearance) of an edible mushroom and a deadly mushroom can be very subtle. So, collecting wild mushrooms for a meal is not a wise idea unless you are an expert mycologist.

9.6 Review Questions

1. In terms of number of species, the most diverse fungi are:

   A. Chytrids

   B. Zygote fungi

   C. Mycorrhizal fungi

   D. Sac fungi

   E. Club fungi

2. All fungi:

   A. Are multicellular.

   B. Have flagellated hyphaes.

C. Have centrioles in their cells.

D. Are heterotrophs.

E. All of the above.

3. Zoospores are produced in:

    A. Zygomycetes

    B. Chytrids

    C. Zytrids

    D. Ascomycetes

4. Coenocytic fungi:

    A. Are not made of cells.

    B. Have cells that collectively form a continuous mass with multiple nuclei.

    C. Have cells separated by cross-walls called septa.

    D. Have cells that never go through cellular division processes such as mitosis.

5. What separates fungi into five phyla?

    A. Reproductive features

    B. Morphology

    C. Means of obtaining nutrition

    D. The ways in which they are useful or harmful

6. The antibiotic penicillin is produced from what kind of fungi?

    A. Zygomycota

    B. Ascomycota

C. Chitridiomycota

D. Lichens

E. Basidiomycota

7. Almost all fungi share all the following characteristics except:

A. They are multicellular.

B. They go through a heterokaryotic stage.

C. They have cell walls made of chitin.

D. They obtain their nutrition by absorption.

E. Their lifecycle is diploid dominant.

8. True or False: Fungi are most closely related to plants and not animals.

9. Which of the following group of fungi has flagellated spores?

A. Ascomycota

B. Basidiomycota

C. Chytridiomycota

D. Glomeromycota

E. Zygomycota.

10. Which of the following is not true about fungi?

A. All fungi are multicellular.

B. All fungi are heterotrophs.

C. Some fungi can grow either as single celled forms or in form of filaments (or hyphae).

D. Some fungi digest the nutrients by secreting digestive enzymes into the environment and then absorb the digested nutrients.

E. A and C

9.7 Answer Key

1.	D	6.	B
2.	D	7.	E
3.	B	8.	False
4.	B	9.	C
5.	A	10.	A

Notes:

Q7. Fungi have a haploid dominant lifecycle.

Q8. Fungi are most closely related to animals.

Chapter 10

Evolution of the Kingdom Animalia

10.1 The Big Picture: What is an Animal Anyway?

Like plants and fungi, animals are eukaryotes. They are multicelled heterotrophs. Examples of some animals are presented in Figure 10.1. The criteria that distinguish animals from plants and fungi are listed in Table 10.1. Although there are always exceptions, the differences listed are mostly true.

There are a few common features that distinguish an organism as an animal:

Cellularity: Animals are multicellular. The cells of animals do not have walls, but instead a number of proteins within the extracellular environment. These proteins, such as collagen, provide the needed structural support for the cells.

Domain: Animals belong to Domain Eukarya.

Nutrition mode: Animals are heterotrophic and ingest their food, as their bodies cannot self-produce all the organic molecules that they need. They are consumers contributing to all three basic ecological roles: detrivores (feeding on dead organisms), parasites (living on or within a host organism and nourishing from it without killing it), and predators

Biology of Evolution and Systematics, by Paul Sanghera
Copyright © 2015 Infonential.

(eating other organisms or parts of them. There are three kinds of predators: carnivores, which eat other animals; herbivores, which eat plants; and omnivores, which eat both plants and animals.

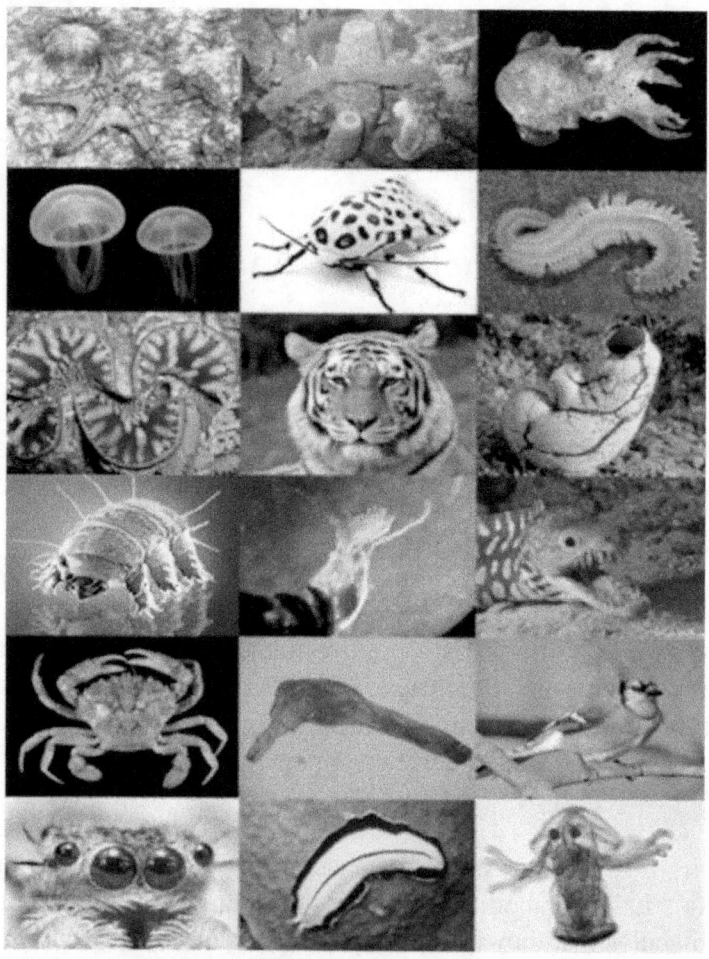

Figure 10.1 The animal kingdom: a few examples of animals showing diversity.

Tissues: Unlike any other groups of organisms, animals have two special types of cells; muscle cells and nerve cells. In most animal groups, these cells are organized into a higher organizational level called, tissues. True tissues develop from embryonic layers called germ layers and there can be up to three of them: endoderm (inner), mesoderm (middle), ectoderm (outer).

> **Note.** Working from the basis for motility and nerve-related functions, muscle and nerve cells are at the heart of many adaptations that diverged animals from fungi and plants.

Reproduction: Animals are diploid, and only their gametes are haploid cells. Unlike the reproduction process of fungi and plants, the animal reproduction process produces haploid reproductive cells (sperms and eggs) directly from meiosis, which fuse to form a diploid zygote that develops into a diploid organism. The diploid stage is a dominate factor in the lifecycle of most animals. Of course, there are some exceptions. For example, some insects, such as bees, may produce haploid offspring, and some animals, such as hydra, may reproduce by budding, an asexual mode of reproduction.

Development: Even though some animals, including humans, develop directly into adults, most animals go through a developmental stage called larva. A larva is a reproductively immature form of an organism that is sharply distinct from the adult form in many aspects, such as habitat, morphology, and nutrition. A larva goes through *metamorphosis,* which is a stage that transforms the larva into either an adult or a juvenile that looks like an adult and then subsequently grows into one.

> **Fascinating Fact!** Although biologists have identified 1.3 million species of animals living on Earth, which in itself is an incredible show of diversity, the actual number of species is much larger and over ninety-nine percent are already extinct.

Biology of Evolution and Systematics

Table 10.1 Comparison of animals, fungi, and plants.

Characteristics	Animals	Fungi	Plants
Mode of nutrition	Heterotrophs: ingests food and digests it inside the body	Heterotrophs: digests food outside the body by releasing enzymes, and then absorbing it	Autotrophs: makes food through photosynthesis
Cellularity	Multicellular	Mostly multicellular Some unicellular	Multicellular
Cell wall	No cell wall: proteins such as collagen provide structural support to cells	Cell walls made of chitin, no collagen	Cell walls: largely composed of cellulose, no collagen
Tissue type	Many animals have muscle cells and nerve cells organized into tissues	Lack muscle cells and nerve cells	Three tissue types: dermal, ground, and vascular
Lifecycle	Diploid stage dominates Only gametes are haploids: sperms and eggs	Haploid stage dominates	Alternation of haploid and diploid generations Haploid generation dominant in non-vascular plants Diploid generation dominant in vascular plants

10.2 Morphological Diversity

The overwhelming diversity among animals can be explained via a relatively small number of basic body plans that have evolved over time. The set of variables that define these various body plans include body symmetry, tissues, body cavities, segmentation, and digestive system.

Symmetry: Symmetry refers to the spatial balance in the distribution of duplicate body parts or regions. With the symmetry criterion, animals fall into three groups: those who do not have any symmetry and are called asymmetric, such as sponges; those who have two-sided symmetry, known as bilateral symmetry, such as humans; and those who have a 360 degree symmetry around one axis, radial symmetry, such as jellyfish. Consistent with the principle of *form fits the function*, animals with radial symmetry are either *sessile*, attached to a substrate, or they are *planktonic*, swimming with the least effort or just drifting. In contrast, animals with bilateral symmetry are actively motile; they move from place to place actively.

Tissues: This variable is one of the components responsible for the enormous diversity found within the millions of animal species on earth, alive and extinct. Overall, animals can be organized into two groups: those that lack true tissues such as sponges, and those that develop true tissues such as humans. A true tissue is a collection of one kind of specialized cell, separated from other tissues by membranes. True tissues develop in general from three embryonic layers called germ layers:

- Ectoderm layer: This is the germ layer that covers the outer surface of the embryo. It gives rise to the covering of the organism and to the nervous system in some groups of animals.

- Endoderm layer: This is the innermost germ layer that gives rise to the lining of the digestive tract, and some organs such as liver and lungs in vertebrates (animals with backbones).

- Mesoderm layer: This, if it is present, is the germ layer in the middle of ectoderm and endoderm layers. This layer forms most of the organs between the digestive tract and also the outer casing of the life form, such as the muscles.

All animals with bilateral symmetry have all three germ layers, and are therefore referred to as *triploblastic*. Animals that have only ectoderm and endoderm layers, such as cnidarians (e.g. jellyfish, corals), are referred to as *diploblastic*.

Body Cavity: A body cavity is any space in the body filled with fluid or air. Most triploblastic animals have a body cavity, called *coelom*, located between the outer body wall and the digestive tract. A coelom is called a true coelom if it is lined by tissues only derived from the mesoderm germ layer. A body cavity derived from both mesoderm and endoderm layers, is called a pseudocoelom. Here, by coelom we mean true coelom, unless otherwise stated. Organs formed inside the coelom can grow and move freely, and the cavity fluid protects them from shocks. Some animals have no coelom, and others have a pseudocoelom or true coelom.

Segmentation: Segmentation is a series or repetition of the same or similar units along an organism's body from the back to the front axis. Repetition allows an organism to specialize in a task. Some of the mesoderm layer in the embryos of certain species, such as humans, forms a series of segments along the body axis, which subsequently develop into different organs or organ systems such as the backbone, the ribs, or the back muscles in the human body. Some animals are segmented, others are not.

Digestive system: A digestive system is an organ system that breaks down food into molecules small enough for the body to absorb it. As animals are heterotrophs, they need to ingest and digest food. Most animals have a mouth to ingest food, but there are many different digestive systems. Some have a full digestive tract, others do not. There are two broad categories of digestive systems in animals. Many invertebrates have a sac-like digestive system with only one opening that is used for both ingesting food and pushing out waste. Other animals, mainly vertebrates, have a more efficient and sophisticated digestive system, which has separate openings for food to enter and for waste to exit.

Within the small number of basic body plans defined by these elements, there are numerous variations in the details that have evolved, producing a great amount of diversity in animal species over time. Recently unearthed fossils suggest that the first animals appeared over 760 million years ago

10.3 Evolution of Animals

Figure 10.2 presents a phylogenetic tree for the major groups (or phyla) of animals. It is largely based on molecular studies. Morphological, cellular, and molecular studies indicate that animals evolved from a colonial flagellated protist that was most likely similar to an existing group of protists called choanoflagellates, which are the closest living relatives of animals. Some of the clades in the figure are described below:

Biology of Evolution and Systematics

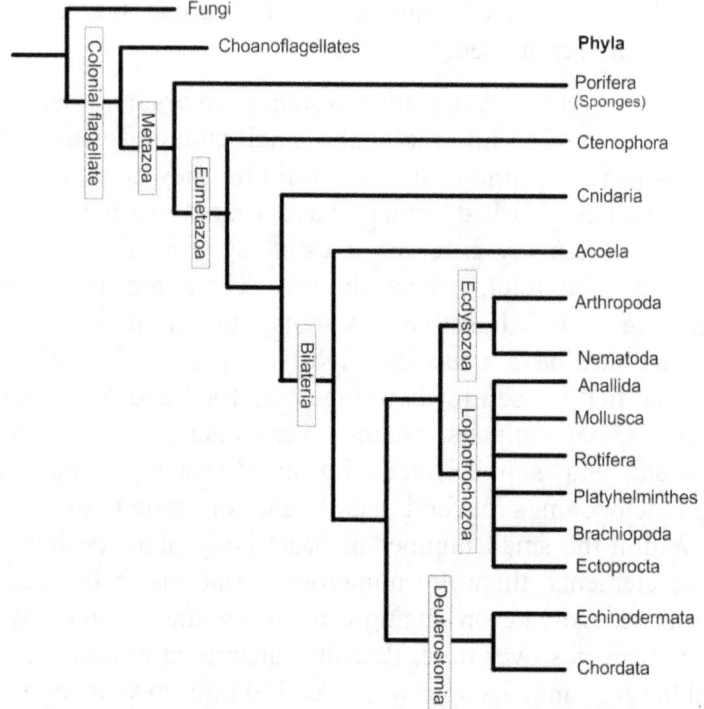

Figure 10.2 An evolutionary tree for major groups of animals, largely based on the molecular data.

Metazoa: This clade is comprised of all animals.

Eumetazoa: Animals that have true tissues belong to this clade, which is almost all except sponges and a few other animal groups,

Bilateria: All animals with bilateral symmetry form this clade.

Ecdysozoa: This is the clade in which animals secrete material that organizes into exoskeletons. Many organisms of this clade regularly shed their exoskeleton to replace it with the new one.

174

Lephotrochozoa: Two groups of animals that support two distinctive features are members of this clade: One has lophophore and the other has trochophore larva. Lophophore is a distinctive appendage that bears tentacles and is used for feeding, and trochophore larva is a distinctive larva stage that a group of animals go through.

Deuterostomia: This is the clade in which the anus is the first thing to form in the embryo. It includes any of the major groups that are defined by this type of embryonic development. This clade includes echinodermata, and chordata, the phylum that we belong to.

Some of the animal phyla shown in Figure 10.2 are described in the following.

Phylum Porifera: These are the organisms with the common name, *sponges*. These animals are without tissues and without any definite body plan. They are aquatic animals and while some sponge species live in fresh water, most of them are marine animals. They are *sessile*, that is, they are attached to some surface, and they get their energy through filter-feeding; most of them feed on bacteria that they filter from water. They are very much like a colony of choanoflagellates with additional types of cells, creating more division of labor among the cells. Some flagellated cells, called *choanocytes*, support water feeding by using their flagella to keep water moving in and out of the sponge's body. Water moves in through pores in the sides of the sponge's body and moves out through a large opening at the top called an *osculum*. Special cells called *amoebocytes* inside the sponge body secrete material that organizes into hard and sharp crystal-like structures called *spicules*, which help the sponge stand upright.

Phylum Cnidaria: This animal phylum includes hydra, jellyfishes, corals, and sea anemones. Like porifera, most of

175

them are marine animals. But unlike porifera, they have true tissues; two layers of germ tissue (ectoderm and endoderm), gastrovascular cavity, and radial symmetry. The gastrovascular cavity has only one opening, and it is used to take in food and to push out waste or undigested food. Cnidarians are predators, and they capture their prey by using tentacles surrounding their mouth. The tentacles have specialized stinging cells known as *cnidocytes*, which discharge structures called *nematocytes* to attack prey. The body plan of cnidarians has two kinds of forms: *polyp* and *medusa*. Medusa cnidarians are free to swim, albeit rather slowly, or to float along with their mouth and tentacles pointing downward. For example, the class scyphozoa, which includes the jellyfish, live mostly as medusas. Quite the opposite, polyp cnidarians stay attached to the substrate, with their mouth and tentacles pointing upward. The class anthozoa, which includes anemones and polyps, lives mostly as polyps. Class Hydrozoa, in which hydra belong, live part of their lives as a medusa and part as a polyp.

Phylum Platyhelminthes: These animals are essentially the flatworms, and they are free living, that is, not attached. Examples are flukes and tapeworms. They have all three germ layers (ectoderm, mesoderm, and endoderm), unlike cnidarians, which have only two germ layers with mesoderm missing. Also their body plan is based on bilateral symmetry, rather than radial symmetry found in cnidarians. Like cnidarians, they have a gastrovascular cavity (with only one opening), although it is branched. However, they are solid flesh with no coelom or body cavity. They are not segmented either. They are the simplest triploblastic animals that have bilateral symmetry and true tissues and organs.

Phylum Rotifera: These are microscopic but multicellular aquatic animals with body cavities. They can be mostly found in freshwater environments, but some species can also be found in damp soil and salt water. They have a complete digestive tract (digestive tube), called an alimentary canal, with separate openings for taking in food (mouth) and for eliminating waste (anus). They have long coronal cilia surrounding their mouth that they use to catch and direct food into their mouth by generating water currents. The role of males in rotifer is either non-existent or very limited. Some rotifer species are only females, which produce offspring from unfertilized eggs, a process called *parthenogenesis*. In other species, females produce two types of eggs. One type will only produce degenerate males, which will die after producing sperms. These sperms are then used to fertilize the other type of eggs to produce females.

Phylum Annelida: These are segmented worms, such as earthworms and leeches. Segmented means made up of repeating units. They have a body cavity that is a true coelom.

Phylum Mollusca: This is a very diverse group of animals, which include chitons, clams, mussels, octopuses, oysters, slugs, snails, and squids. Most of the species are marine animals, but some species have freshwater and terrestrial habitats. They have bilateral symmetry and segmented bodies. They form a clade in which all the members typically having a muscular foot, a visceral mass, and a mantle (a hard exoskeleton made of calcium carbonate). They vary in size from a microscopic, a millimeter or less in length, such as aplacophorans, to a few meters long organisms, such as twenty meter long squids.

Nematoda: These are roundworms, cylindrical and not segmented. They have an internal cavity with tapered ends, which is a pseudocoelom, that is, not a true coelom. They lack cilia or a well-defined head. Most of the species are decomposers in soil and water, although some of them are parasites to other animals and plants.

> **Note.** Arthropods (such as crabs, insects, and spiders) sand nematodes (such as roundworms) are among the most abundant of all animal groups.

Phylum Arthropoda: This group includes crabs, crayfish, crustaceans, insects, lobsters, scorpions, shrimps, and spiders. The key characteristics of this phylum include exoskeletons, jointed appendages, and segmentation. Arthropods are unquestionably considered the most successful of all phyla, and they owe their success to their key characteristics. The number of species of insects alone is estimated in the tens of millions. They can be found everywhere: land, sea, and air. Their exoskeleton is formed from stiff cuticles, which are largely made of chitin and proteins. In some cases, the exoskeletons are further hardened with calcium carbonate. The jointed appendages may be modified in different ways to form different structures such as antennae, mouthparts, and reproductive organs.

> **Note.** In biology, the success of a group of organisms is measured by their numbers and their diversity. Insects alone, all of which have three pairs of legs and one or two pairs of wings, are the most diverse group of organisms, means they have the largest numbers of species, and thereby dominate the terrestrial habitats.

Phylum Echinodermata: The beachgoers or beachcombers are familiar with this phylum, as it includes starfish, brittle stars, sea urchins, sea lilies, sea cucumbers, and sand dollars. This is the largest phylum of marine animals that has no presence in freshwater or on land. They have an exoskeleton of interlocking spines and plates stiffened with calcium carbonate. Even though most adults in this phylum have a radial symmetry, the larvae are bilateral. This is explained in DNA sequencing studies, which indicate that radial traits evolved secondarily in this evolutionary lineage. Echinoderms have a decentralized nervous system, which responds to danger, food, and sexual mates from any direction. However, they have no brain. Although some sea lilies are attached to the seafloor, most adult echinoderms glide around via their little fluid-filled, tube-shaped feet, which are part of their unique water-vascular system. They also use these handy feet to grip objects. All echinoderms have an internal support system called an endoskeleton. Chordata is the closest evolutionary relative to the phylum Echinodermata.

Phylum Chordata: This phylum should be the most familiar to you as it includes you and all other vertebrates, the animals with a backbone, although chordata also includes some invertebrates. All chordates have the following features at some stage of their life:

1. Dorsal nerve cord: A hollow verve cord comprised of a bundle of nerve fibers that run down the back and connect the brain with many of the organs of the body such as the lateral muscles.

2. Notochord: A cartilaginous rod of stiff but flexible connective tissues that support the nerve cord and that run underneath it. In other words, it extends the length of the body and supports the body.

3. Pharyngeal slits: A series of openings across the walls of the pharynx (throat region), which connect the outside of the neck to the inside of the throat. In many cases, these are used as gills.

4. Post-anal tail: This muscular extension of the body is just past the anal opening.

For many vertebrates such as human, these traits may only be present in the embryo.

Fantastic Fact! Dare to compare a fly with a human: you won't see much similarity as their evolutionary lineages diverged hundreds of millions of years ago. However, scientists have found that similar sets of conserved genes govern their development; another example of unity behind diversity and more evidence of evolution occurring.

In Figure 10.2, you will note that only the phylum Chordata includes vertebrates. This is representative of the fact that about ninety-five percent of animal species are invertebrates.

10.4 More Definitions of Terms and Concepts

In this section, we define those important terms and concepts related to the topic at hand that are not already discussed in this chapter.

Homeotic genes: A type of regulatory genes that regulate or control the spatial dimensions of body parts, for example where in the body to place the parts, in animals, fungi, or plants.

Homeobox: A specific 180-nucleotide long DNA sequence, which is found among homeotic genes and also among particular developmental genes that are highly conserved in different groups of animals.

Hox genes: These are homeotic genes in animals. They are called hox (short for *containing homeobox*), because the homebox sequence was first discovered in these genes.

Cleavage: Cleavage is an early stage of development in animals. It includes the process of cytokinesis, which divides the cytoplasm to form two separate daughter cells from one parent cell right after a cell division process, such as mitosis, meiosis I, or meiosis II. During cleavage, a rapid cell division converts a zygote into a ball of cells.

Determinate cleavage: The cleavage is called determinate when it determines the development path of embryonic cells very early in the process. This means that after this determination, an embryonic cell, if separated, cannot develop into a full embryo, and hence cannot develop into a full organism.

Indeterminate cleavage: This is the type of cleavage in which each cell produced during the early cleavage process retains the capability to develop into a complete embryo and therefore a complete organism. The identical twins in humans are made possible by the indeterminate cleavage. Echinodermata and Chordata have indeterminate cleavage.

Sessile: A living animal attached to a substrate.

Caution! Sponges lack hox genes. Instead, other homeobox genes influence their shape.

10.5 In a Nutshell

Molecular studies indicate that animals, the multicellular heterotrophs, originated from a colonial flagellated protist that was similar to modern choanoflagellates. Animals can be broadly categorized into two groups: vertebrates, which have a backbone; and invertebrates, which do not have a backbone. Invertebrates comprise about ninety-five percent of all animal species. All vertebrates belong to the phylum Chordata, which also includes some invertebrates. Sponges comprise a phylum called Porifera, do not have true tissues, and are basal animals on the evolutionary tree. Flatworms are the simplest animals with three germ layers, bilateral symmetry, and true tissues and organs. Most animals reproduce sexually, and their lifecycle is dominated by the diploid stage: organisms are diploid and gametes are haploid.

10.6 Review Questions

1. The common ancestor of all animals was a:

    A. Choanoflagellate

    B. Bacteria

    C. Colonial flagellated protist

    D. Fungus

2. Animals with bilateral symmetry are:

    A. Diploblastic

    B. Triploblastic

    C. Monoblastic

    D. Herbivores

3. The structurally simplest animals with a three-layered embryo are:

    A. Flatworms

    B. Jellyfish

    C. Polyps

D. Porifera

4. Where do echinoderms live?

   A. Marine water

   B. Fresh water

   C. Terrestrial

   D. A and B

5. A water vascular system with tube feet is a unique trait of:

   A. Sponges

   B. Starfish

   C. Flukes

   D. A and B

6. What cells produce spicules, which help the sponge to stand upright?

   A. Osculum

   B. Choanocytes

   C. Sessile

   D. Amoebocytes

7. The four defining features of Chordata are:

   A. Dorsal nerve cord, notochord, backbone, and pharynx with slits.

   B. Dorsal nerve cord, notochord, tail extending past the anus, and pharynx with slits.

   C. Dorsal nerve cord, notochord, head, and pharynx with slits.

D. Dorsal nerve cord, notochord, tail extending past the anus, and coelom.

8. Which animal phylum has a unique stinging structure?

   A. Cnidaria

   B. Porifera

   C. Rotifera

   D. Mollusca

9. Which feature in reproduction is responsible for identical twins in human?

   A. Determinate cleavage

   B. Indeterminate cleavage

   C. Radial cleavage

   D. Bilateral cleavage

10. Which of the following is an incorrect statement?

   A. All chordates have segmented body plan.

   B. Some rotifers reproduce from unfertilized eggs.

   C. Chordata contains all the vertebrates, whereas other phyla contain all the invertebrates.

   D. Cnidarians have only two germ layers: ectoderm and endoderm.

10.7 Answer Key

1. C	6. D
2. B	7. B
3. A	8. A
4. A	9. B
5. B	10. C

Notes:

Q2. Animals with bilateral symmetry have a mesoderm layer in addition to ectoderm and endoderm layers.

Q5. A water vascular system, with tube feet, is a unique trait of echinoderms, and a starfish is an echinoderm.

Q8. Cnidarians have unique stinging structures known as nematocysts that are contained in specialized cells called cnidocytes.

Q10. Chordata also contains some invertebrates. All vertebrates are chordates, but not all the chordates are vertebrates.

Biology of Evolution and Systematics

Glossary

adaptation An inherited trait of an organism that enhances the capability of the organism to survive and reproduce viable offspring in a given environment.

adaptive evolution Evolution that improves the fit between a population and its environment.

adaptive radiation The process in which a lineage in the evolutionary tree rapidly gives rise to many new species through adaptations, which enable the new species to fill different niches offered by the environment.

alleles Different versions of a gene.

allopatric speciation The formation of new species due to a geographic separation between two populations of the same species.

analogy Similarity in inherited traits between different species resulting from convergent evolution not as much from shared ancestry.

artificial evolution The evolution that is carried out by humans by using non-natural (artificial) means.

allopolyploidy The polyploidy that occurs between organisms of different species.

apomorphy Any derived states or features in a taxon; can be autopomorphy or synapomorphy.

autopolyploidy The polyploidy that occurs between organisms of the same species.

autapomorphy Any derived feature that is unique to a single taxon.

balancing selection The form of natural selection that maintains multiple forms of a phenotype, called balanced polymorphism, in a population.

binomial A formula for naming a species: the genus with first letter uppercased followed by the species (specific epithet), both italicized. Example: *Homo sapiens* for humans.

biofilm A colony of prokaryotes coating a surface. It is formed by prokaryotes by adhering to one another.

biogeography The study of patterns of the geographical distribution of species and communities in the present and in the past.

body cavity Any space in the body filled with fluid or air.

bottleneck effect The genetic drift that occurs as a result of a sudden event in the environment that reduces population size. The population after the event has a different gene pool than the population before the event.

Cambrian explosion Occuring around 530 million years ago, this was was a relatively brief geological time, a few million years, during which a burst of evolutionary changes gave rise to many phyla of animals that exist today and diversified phtoplanktons and calcimicrobes.

chemoautotroph Any organism that uses some form of carbon dioxide as a carbon source and inorganic chemicals such as sulfur, hydrogen sulfide, ammonia, and ferrous iron as an energy source.

chemoheterotroph Any organism that uses organic compounds such as glucose as a carbon source and as an energy source as well.

clade A group of taxa in a cladogram that includes one ancestral taxon and all of its descendents.

cladistics A phylogenetic approach in systematics that uses common ancestry to classify taxa of organisms into groups called clades.

Cladogram An evolutionary tree built according to the rules of cladistics in which each node (branch point) represents a speciation event wherein an ancestral species splits into two daughter species.

class A group of closely related orders in the taxonomic system.

coelom A body cavity present in most triploblastic animals, lined by tissues only derived from the germ layer mesoderm.

coevolution The process through which two species evolve together by exerting evolutionary pressure on each other.

conjugation A genetic recombination process in which one prokaryote transfers some of its DNA into another prokaryote directly through a structure called a sex pilus.

convergent evolution The evolution of similar features among independent lineages of organisms due to the similarities in their environments at the respective places where they evolved.

Cretaceous mass extinction The mass extinction event that occurred about 65 million years ago and wiped out many (eighty-five percent) species including dinosaurs. Also called the K-T extinction or K/T extinction.

diploblastic Any organism that contains only two germ layers: endoderm and ectoderm.

diploidy The characteristics of organisms whose cells have two sets (2n) of chromosomes, one set inherited from each parent.

dispersal The process that moves one or more parts of a population to new geographic locations separated from the original location. *See variance.*

domain The highest taxonomic group or rank. All organisms are organized into three domains: Archaea, Bacteria, and Eukarya.

endosymbiosis A process in which a unicellular organism engulfs another unicellular organism, which over time turns into an organelle within the host organism.

eukaryotes Organisms made of eukaryotic cells, the cells that contain membrane-bound internal structures called organelles, such as a nucleus.

evolution Emergence of lines of descent with modifications from common ancestors over generations of organisms.

exoenzymes The digestive enzymes that fungi secrete to their external environment in order to digest food before they absorb it.

family A group of closely related genera in the taxonomic system.

F factor The segment of a DNA in a prokaryote that enables it to form a sex pilus used by the prokaryote to transfer genes to another cell.

founder effect The genetic drift that occurs when a few organisms depart from a large population and make their own population.

frequency dependent selection The natural selection in which the relative reproduction fitness of a phenotype depends on the relative number (frequency) of the phenotype in the population, that is, how common the phenotype is.

genetic drift The process in which the events based on chance or probability cause changes in allele frequencies of the population's gene pool from one generation to the next.

genus A group of closely related species in the taxonomic system.

gene flow The transfer of genes or alleles into a population or out of the population, which can occur due to multiple reasons such as immigration and emigration.

gene pool The sum total of all the alleles (versions) of all the genes of all the organisms of a population.

genotype Genetic makeup; set of alleles of an organism.

heterozygote advantage The natural selection in which heterozygous organisms in a population have a greater relative fitness than homozygous organisms.

homology Similarity in inherited traits, such as body parts, between two species resulting from a shared ancestry.

horizontal gene transfer Gene transfer between the genome of one organism to the genome of another organism, when the two organisms do not have a parent-offspring relationship.

Biology of Evolution and Systematic

hybrid Offspring that is produced as a result of mating between organisms from two different species. Also called *interspecific hybrids.*

hyphae A filament of the filamentous structure of a fungus.

inbreeding A type of nonrandom mating in which members of a population only mate with their close relatives.

intersexual selection The sexual selection in which an organism exerts selective pressure on other organisms of the opposite sex, for instance, by being selective in choosing a mate.

intrasexual selection The sexual selection in which an organism exerts selective pressure on other organisms of the same sex.

kingdom A group of closely related phyla in the taxonomic system. The second broadest taxonomic category, second only to domain.

macroevolution Evolution recorded or observed at the species and higher levels. Examples: appearance of new species or groups of organisms, extinction of species, change in diversity of life due to mass extinction.

megaphyll A large leaf with a network of veins.

metabolism The sum total of all biochemical reactions that cells use to manage their energy resources.

microevolution The evolution of the gene pool of a population in terms of changes in allele frequencies.

microphyll A small and, in most cases, a spine-shaped leaf with a single unbranched vein.

molecular clock A technique in evolutionary science which estimates how much time an evolutionary change took, by using the fact that some regions of the genome change at a constant or known rate.

monophyletic group A group of taxa that includes one ancestral taxon and all its descendents. A monophyletic group is a clade.

Monera The kingdom of unicellular organisms without a nucleus or organelles in their cells, that is, prokaryotes.

morphological species A group or groups of organisms characterized by the unique set of measurable anatomical features such as body shape.

mutation A change in the nucleotide sequence of DNA that may arise for various reasons such as errors in DNA replication.

mycelium The whole collection or network of hyphae in a fungus.

mycorrhizae The fungus roots that represent a mutualistic symbiotic relationship between a host plant and the fungi that colonize the host plant's roots.

natural selection A process that allows organisms with a certain set of traits to survive and reproduce offspring more viable than organisms that do not have these traits.

order A group of closely related families in the taxonomic system.

paleontotlogy The study of fossils: the preserved remains or traces of organisms that lived in the past.

paraphyletic group A group of taxa that consists of one ancestor and some (not all) of its descendents.

193

Biology of Evolution and Systematic

Permian mass extinction The biggest mass extinction event in the history of life on Earth that occurred about 251 million years ago. It wiped out many species including ninety-six percent of marine animal species and seventy percent of terrestrial vertebrate species.

phenotype The visible physical and physiological characteristics (traits) of an organism. Phenotypes arise from corresponding geneotypes.

photoautotroph Any organism that uses some form of carbon dioxide as a carbon source and light such as sunlight as an energy source. Photoautotrophs carry out photosynthesis.

photoheterotroph Any organism that uses organic compounds, such as glucose as the carbon source and sunlight as an energy source. Photoheterotrophs carry out photosynthesis.

phylogenetic species The smallest group of organisms that share a common ancestor and is represented by a branch on the evolutionary tree of life.

phylogenetics The study of phylogeny.

phylogeny The evolutionary history of a species or groups of species.

phylum A group of closely related classes in the taxonomic system.

Plesiomorphy Any characteristic of a taxon in its ancestral state. When a characteristic evolves from one state to another, the original (ancestral) state is called plesiomorphy.

polyphyletic group A group of taxa that does not have a common ancestor but do have at least one common or similar characteristic shared by all the taxa in the group. The common characteristics may have appeared through convergent or parallel evolution.

polyploidy The process in which an organism's genome acquires more than two sets of chromosomes.

population A group of organisms of the same species living in the same area at the same time.

prokaryote Any single celled organism made out of a prokaryotic cell, which is a cell that has no nucleus or any other organelles (membrane enclosed internal structures).

protocell A primitive version of a living cell. Protocells contained an internal biochemical environment different from the external environment, and they were protected by a simple fatty acid membrane.

punctuated equilibrium The sudden evolutionary change observed in fossil records that is followed or preceded by a relatively static period with no apparent change.

relative fitness The contribution that an organism makes to the gene pool of the next generation as compared to the contributions of other organisms in the population.

reproductive isolation Any of the barriers that prevent or impede organisms from two populations of the same species or from two different species, from interbreeding and producing viable and fertile offspring.

rhizoid A tube-shaped (hair-like) cell or a filament of a cell that anchors bryophyte plants to the ground.

septa The cross-walls that separate fungi cells from one another.

sexual dimorphism The differences between males and females in secondary sexual traits such as behavior, color, ornamentation, and size.

sexual selection A type of natural selection that operates on a set of traits directly involved in obtaining mates; organisms with certain inherited traits have an advantage in attracting mates over those organisms that lack these traits.

sorus A little brown dot under the leaf of a fern; this structure contains dozens of sporangia.

speciation The rise of a new species.

speciation event An evolutionary event in which one species splits into two or more species due to modifications that evolved over generations.

species One or more groups of organisms that have the potential to interbreed and produce viable and fertile offspring but do not have the same potential to interbreed and reproduce with members of other groups. Also called *biological species.*

sporangium An organ in plants in which meiosis occurs to produce haploid cells called spores.

sporophyll A modified leaf that bears sporangia and is therefore a reproductive part of a plant.

sympatric speciation The process in which a new species forms within the same population without a geographic split.

synapomorphy Any derived feature that is shared by two or more taxa, which thus helps to identify their clade.

systematics A sub-field of biology focused on the study of biodiversity via classifying organisms and determining their evolutionary relationships.

taxon   A group of organisms defined at any level of taxonomy or classification, from species through kingdom.

taxonomy   The system developed by Swedish physician, Carolus Linnaeus to classify the diverse forms of life into hierarchical groups called species, genus, family, order, class, phylum, and kingdom.

transformation   A genetic recombination process in which a prokaryote takes in foreign DNA from its surrounding.

transduction   A genetic recombination process in which a virus injects foreign DNA into a prokaryote.

triploblastic   Any organism that contains three germ layers: endoderm, mesoderm, and ectoderm.

variance   The process in which a geological range that houses a population splits into two disjoint parts, creating a barrier for gene flow between the two segments of the population in the two separate parts.

vestigial structure   A useless remnant of a structure in a species that was once useful in the ancestral species but lost all or most of its functions during the course of evolution.

Biology of Evolution and Systematic

Credits and Acknowledgments

Unless otherwise acknowledged, all pictures and illustrations in this book are the property of Infonential, Inc. We have made our best effort to trace and acknowledge the ownership of the following items. In the event of any question or issue arising from the use of one of these items, we will be pleased to make the necessary corrections in future printings.

Figure 1.3 Courtesy of Tanya Bonakdar Gallery, New York.

Figure 1.4 Courtesy of University of Michigan Museum of Zoology, website.

Figure 3.3 Courtesy of E.A. Kellogg and the U.S. National Academy of Sciences.

Figure 3.4 GNU Free Documentation License.

Figure 4.2 Public domain.

Figure 4.4 Credit: Understanding Evolution. 2012. University of California Museum of Paleontology, 9 February 2012 evolution.berkeley.edu., U.S. National Center for Science Education, and U.S. National Science Foundation.

Figure 6.3 Courtesy of Mariana Ruiz Villarreal.

Figure 7.1 GNU Free Documentation License.

Figure 9.1 Courtesy of Dr. David Midgley.

Figure 9.2 Public domain.

Figure 10.1 GNU Free Documentation License.

Biology of Evolution and Systematics